# Topics in Food and Drug Law

## HOW TO COMPLY WITH DRUG CGMPs

*Cathy L. Burgess*
*Daniel G. Jarcho*

FDLI

 ISBN 978-1-935065-83-8

For additional information or to order copies of this publication, contact FDLI:

FDLI

1155 15th Street NW, Suite 910
Washington, DC 20005
(800) 956-6293 or (202) 371-1420
info@fdli.org
www.fdli.org

# Table of Contents

## Table of Contents, continued

# About FDLI

The Food and Drug Law Institute (FDLI), founded in 1949, is a nonprofit membership organization that offers education, training, publications, and professional engagement opportunities in the field of food and drug law. As a neutral convener, FDLI provides a venue for stakeholders to inform innovative public policy, law, and regulation.

FDLI's scope covers all industries regulated by the U.S. Food and Drug Administration (FDA) and related agencies and authorities in the U.S. and globally.

## Disclaimer

This publication is intended to be informational and does not constitute legal advice regarding any specific situation.

# About the Authors

**Cathy L. Burgess**

Cathy L. Burgess is a Partner in the Health Care Group of Alston & Bird LLP in Washington, DC. FDA-regulated entities benefit from Ms. Burgess's common-sense advice for achieving business objectives that avoid compliance and enforcement risks. With more than 25 years of extensive experience in the areas of CGMP regulation and product risk management, she provides strategic counseling, and works closely with clients to identify and address potential risks throughout the product lifecycle. Ms. Burgess advises clients on issues related to management responsibility, supply chain management, complaint handling, recalls, and inspection management. She assists manufacturers in responding to Form FDA-483s and Warning Letters, and has developed compliance plans and remediation strategies, working closely with clients and technical consultants. She regularly provides training on FDA compliance and enforcement issues, and serves on the Food and Drug Law Institute's Drugs and Biologics Committee and Audit Committee. Prior to joining the firm, Ms. Burgess served as associate general counsel for the American Red Cross, responsible for regulatory matters. In this role she provided legal assistance and strategic advice to Red Cross senior management and the Board of Governors' Audit Committee on matters related to the Red Cross Amended Consent Decree. Ms. Burgess served as the defense team's first chair for expert testimony on CGMPs and analytical method validation in United States v. Barr Laboratories.

**Daniel G. Jarcho**

Daniel G. Jarcho is a Partner in the Litigation and Trial Practice Group of Alston & Bird LLP in Washington, DC. Dan Jarcho's practice focuses on litigation involving the Food and Drug Administration (FDA) and the industries that FDA regulates. He also counsels clients about resolving regulatory compliance disputes with FDA that could lead to litigation, including disputes over inspections, Form FDA-483 observations, Warning Letters, recalls, and import alerts. Mr. Jarcho has represented clients in numerous FDA-related civil actions for seizures, injunctions, and civil penalties and in a wide variety of criminal investigations initiated by FDA. He also has significant experience representing plaintiffs in lawsuits challenging FDA regulatory actions under the Administrative Procedure Act. Mr. Jarcho also has litigated the federal preemption defense in a large number of product liability lawsuits involving products regulated by FDA. He litigates cases in both the federal trial courts and the federal appellate courts. Before joining the firm, Mr. Jarcho was a Trial Attorney in the U.S. Department of Justice's Office of Consumer Litigation, where he represented FDA in federal court civil and criminal litigation, including enforcement actions for seizure and injunction, Administrative Procedure Act cases, and grand jury investigations.

**Acknowledgments**

The authors would like to acknowledge Alston & Bird associates Aaron Block, Brendan Carroll, Justin Mann, Jenny Mendelsohn, and Elise Paeffgen, who assisted in editing, cite checking, and proofreading this publication. We are truly grateful for the dedication, hard work, and sense of humor.

# Introduction

It has been almost 25 years since the U. S. District Court for New Jersey issued its opinion in *United States v. Barr Laboratories, Inc.*, 812 F. Supp. 458, (D.N.J. 1993), widely recognized as the landmark case in the area of current Good Manufacturing Practice (CGMP) regulation. The *Barr* case was significant because it changed the way that the Food and Drug Administration (FDA) and industry began to think about "current" standards for good manufacturing practices. As the opinion illustrates, prior to the *Barr* litigation, FDA's view was that CGMPs were practices employed by "better" firms.[1] Barr's view was that there were many ways to comply with CGMPs: as long as a firm's approach was rational and did not violate general principles of CGMPs, that was acceptable for purposes of compliance.[2]

The presiding judge, Judge Alfred Wolin, was clearly frustrated by the lack of well-articulated standards, stating in the opinion that the lack of clear standards had resulted in "an industry mired in uncertainty and conflict, guided by vague regulations which produce tugs-of-war of varying intensity."[3] Whether or not the judge intended to establish industry-wide standards through the *Barr* opinion, that is exactly what happened. As described in greater detail in this primer, FDA investigators began to carry copies of the opinion with them during FDA inspections. In addition, FDA's thinking on certain issues, particularly with respect to out-of-specification investigations, was influenced by the opinion. Some of the flaws in the opinion (e.g., the number of retests required as part of an investigation, "7 out of 8 testing")[4] forced FDA to develop new guidance that reflected an evolving, science-based approach to CGMPs for pharmaceutical products. Moreover, given the court's holding that firms should look to scientific literature if FDA's regulations or guidance were ambiguous, FDA began to work diligently to remove ambiguity and provide uniform standards for compliance.

For almost 10 years, CGMP compliance was heavily influenced by the *Barr* decision. In August 2002, FDA launched its "Pharmaceutical Current Good Manufacturing Practices (CGMPs) for the 21st Century" initiative, with the stated goal of "moderniz[ing] the regulation of pharmaceutical manufacturing and product quality" for human and veterinary drugs, as well as certain biological products, such as vaccines.[5] The initiative was designed to encourage adoption of new technologies, quality systems approaches, and science-based and risk-based decision making. Another goal was to enhance quality systems approaches within the agency.

As FDA moved to a new paradigm for regulatory compliance, changes in the pharmaceutical industry continued to outpace FDA's efforts. A variety of factors, such as flat growth in the prescription drug market, the need to reduce healthcare costs, and increasing government involvement in management of prescription drug costs, has forced manufacturers to search for lower-cost suppliers.[6] As the drug manufacturing supply chain

---

1 *United States v. Barr Labs., Inc.*, 812 F. Supp. 458 (D.N.J. 1993).

2 *Id.*

3 *Id.* at 464.

4 *Id.* at 470 n.9.

5 Food & Drug Admin., Pharmaceutical CGMPs for the 21st Century – A Risk-Based Approach Final Report 1 (2004), http://www.fda.gov/downloads/Drugs/DevelopmentApprovalProcess/Manufacturing/QuestionsandAnswersonCurrentGoodManufacturingPracticesCGMPforDrugs/ucm176374.pdf.

6 For example, the cost of active pharmaceutical ingredients is 15-40 percent lower in India than in the United States. Food & Drug Admin., Pathway to Global Product Safety and Quality 10 (2011). In addition, as of 2011 there were approximately 1,400 drug companies in

continued to expand beyond U.S. borders, unanticipated compliance issues emerged: too few FDA investigators in the international inspection cadre to inspect the growing number of regulated firms outside the U.S., language barriers, cultural issues, and economically motivated adulteration. In light of these developments, FDA sought, and recently obtained, new CGMP authority from Congress to address the complex issues associated with globalization of the pharmaceutical industry.

The purpose of this book is to provide members of the legal community and regulated industry a resource that describes the historical and legal bases for, and the governing principles of, CGMPs. It also provides information about current standards for compliance. It is likely that these standards will change over time, but the underlying concepts that we discuss should help the reader understand these changes, and why implementation of "current" Good Manufacturing Practices is essential for regulatory compliance.

China capable of producing more than 60 dosage forms and more than 5,000 types of drug products. At that time, China supplied more than 5.7 percent of the world market, valued at approximately $50 billion. *Id.* This market share grew at a 19.3 percent compound annual growth rate from 2006 to 2011. *Id.* at 1-2.

# Chapter I
# What Are CGMPs?

## A. Principles

Regardless of changes in the Food and Drug Administration's (FDA's) thinking, technological advances, or challenges associated with increasing globalization, the basic principles of current Good Manufacturing Practices (CGMPs) have remained constant. They are as follows:

- Quality, safety, and effectiveness must be designed and built into a product;
- Quality cannot be inspected or tested into a finished product; and
- Each step of the manufacturing process must be controlled to maximize the likelihood that the finished product meets all predetermined specifications.

## B. Legal Framework

The Federal Food, Drug, and Cosmetic Act (FDCA) prohibits introduction of a new drug into interstate commerce unless a New Drug Application (NDA) or Abbreviated New Drug Application (ANDA) has been submitted to, and approved by, FDA.[1] The agency requires each application to include, among other things, "a full description of the methods used in, and the facilities and controls used for, the manufacture, processing and packing of such drug."[2] If FDA determines, after notice and opportunity for a hearing, that the "methods used in, and the facilities and controls used for" drug manufacturing are inadequate, FDA has the authority to withdraw an applicant's NDA.[3]

The FDCA states that a drug product is adulterated if "the methods used in, or the facilities or controls used for, its manufacture, processing, packing or holding do not conform to or are not operated or administered in conformity with current good manufacturing practice to assure that such drug meets the requirements of this chapter as to safety and has the identity and strength, and meets the quality and purity characteristics, which it purports or is represented to possess."[4] The FDCA also provides that the introduction of an adulterated product into interstate commerce is a prohibited act.[5] FDA's CGMP regulations for drugs, set forth at 21 C.F.R. Parts 210 and 211, establish minimum standards for CGMP compliance.

---

1 21 U.S.C. § 355(a).
2 *Id.* § 355(b)(1)(D).
3 *Id.* § 355(e).
4 *Id.* § 351(a)(2)(B).
5 *Id.* § 331(a).

## 1. Drug Amendments of 1962 (Kefauver-Harris Amendments)

The Drug Amendments of 1962 (also known as the Kefauver-Harris Amendments) represented a major expansion of the FDCA, and gave various new authorities to FDA. Under the expanded FDCA provisions, manufacturers were, for the first time, required to demonstrate that their drug products were both safe and effective, obtain FDA approval prior to introducing a new drug, report on adverse events, and follow CGMPs.[6]

The CGMP statutory authority language was considered and modified several times as the Drug Amendments of 1962 were being developed in both House and Senate bills. The various revisions to the CGMP statutory text resulted in an extensive legislative history. On July 19, 1962 the Senate Judiciary Committee Report was issued on the Senate bill version of the Drug Amendment Acts of 1962. The bill had the following statutory language regarding CGMPs:

> [A drug is adulterated] if it is a drug and the methods used in, or facilities or controls used for, its manufacture, processing, packaging, or holding do not conform to current good manufacturing practice to assure that such drug meets the requirements of this Act as to safety and has the identity and strength, and meets the quality and purity characteristics, which it purports or is represented to possess.[7]

The Committee Report's discussion of this section stated that it expanded adulteration under the FDCA to include the failure to comply with CGMPs.[8] The Report then re-emphasized that regulations issued under this provision would be interpretive and prima facie evidence of CGMP standards in a proceeding.[9]

Subsequent to this Report, the Committee received a letter from President Kennedy that contained proposed amendments to the Senate bill.[10] With respect to CGMPs the President's letter proposed to eliminate the text describing the regulations as interpretive, and instead to authorize formal rulemaking.[11] The President's letter expressed concern that interpretive regulations would invite "endless de novo litigation."[12] If CGMPs were interpretive then the regulations only provided "prima facie evidence of adulteration."[13] In other words, a CGMP violation would strongly suggest adulteration under the FDCA but not definitively establish a violation. The Committee ultimately agreed to strike the interpretive aspect of the regulations.[14] However, instead of following the President's request for formal rulemaking, the Committee opted for general or informal rulemaking authority for CGMP regulations under section 701(a) of the FDCA.[15] During the floor debate, Senator Eastland, Chairman of the Senate Judiciary Committee, stated the following regarding CGMP regulations:

---

6 See Drug Amendments of 1962, Pub. L. No. 87-781, § 101-04, 76 Stat. 780, 780-85 (1962).

7 *Id.*

8 S. Rep. No. 87-1744, at 9 (1962).

9 *Id.*

10 Current Good Manufacturing Practice in Manufacture, Processing, Packing, and Holding, 43 Fed. Reg. 45,014, 45,021 (Sept. 29, 1978) (to be codified at 21 C.F.R. pts. 210-11).

11 S. Rep. No. 87-1744, pt. 2, at 3 (1962).

12 *Id.* at 4.

13 Current Good Manufacturing Practice, 43 Fed. Reg. at 45,021.

14 S. Rep. 87-1744, pt. 2, at 4.

15 *Id.*

> [T]he competitive position of responsible manufacturers depends in large part on the confidence of the medical profession and the public, it will be in their own interest to maintain high standards of current good manufacturing practice which will provide a readily determinable basis for enforcement . . . . The Secretary could use his general rulemaking authority under section 701(a) of the act to announce what he, in the administration of the act, considers to be good manufacturing practice . . . .[16]

Senator Kefauver concurred with Senator Eastland's comments, saying "the speech by the distinguished Senator from Mississippi (Mr. Eastland) . . . has accurately defined the provisions of the substitute which was reported. . . . This provision has been strengthened considerably in comparison to the bill which was reported in July [(the previous bill with interpretive CGMP regulations)]."[17] There was a dramatic shift at this point in the evolution of the Drug Amendments of 1962. Not only were CGMP regulations going to be binding in the Senate bill, but they were not going to require lengthy formal rulemaking procedures.

The House during this time was working on their version of the Drug Amendment Bill. As reported out of the Committee on Interstate and Foreign Commerce, the House's version of the CGMP statutory text read as follows:

> [A drug is adulterated] if it is a drug and the method used in, or the facilities or controls used for, its manufacture, processing, packing, or holding do not conform to or are not operated or administered in conformity with current good manufacturing practice (as determined in accordance with regulations promulgated by the Secretary[])to assure that such drug meets the requirements of this Act as to safety and has the identity and strength, and meets the quality and purity characteristics, which it purports or is represented to possess.[18]

In addition, section 101(b) of the House bill would have required CGMP regulations to be issued by formal rulemaking under section 701(e) of the FDCA.[19] During the debate on the House bill, Representative Schenck proposed several amendments including one that would delete the parenthetical reference to regulations promulgated by the Secretary from the CGMP statutory text and eliminate section 101(b) of the House bill.[20] The Schenck amendment was agreed to by the House.[21] As a consequence, the new House bill contained language identical to that in the Senate bill, and thus, no issues were raised in the Conference Committee regarding the CGMP statutory text.[22] The Drug Amendments of 1962 were signed into law on October 10, 1962.[23]

## 2. CGMP Regulations—Implementation of Kefauver-Harris Amendments

Shortly after the passage of the Drug Amendments of 1962, FDA initiated rulemaking for the first set of CGMP regulations. The agency published a set of CGMP rules in 1963, covering 12 subject areas ranging from equipment to laboratory controls.[24] These new regulations allowed for increased enforcement scrutiny over

---

16 108 Cong. Rec. 17,365 (1962).
17 *Id.* at 17,367-68.
18 H.R. Rep. No. 87-2464, at 33 (1962).
19 *Id.* at 56.
20 108 Cong. Rec. 21,056 (1962).
21 *Id.* at 21,096.
22 See H.R. Rep. No. 87-2526 (1962) (Conf. Rep.).
23 Drug Amendments of 1962, Pub. L. No. 87-781, 76 Stat. 780.
24 Drugs; Statement of Ingredients; Prescription Drug Advertisements, 28 Fed. Reg. 6375, 6385 (June 20, 1963) (to be codified at 21 C.F.R. pt. 133).

specific drug manufacturing processes as investigators now could reference codified standards.[25] Through the 1960s and 1970s FDA would, on occasion, revise and expand its CGMP regulations.[26]

In 1976, FDA determined that its existing CGMP regulations for drug manufacturing were in need of significant revisions to reflect technological developments.[27] In addition, other government agencies proposed revisions to the CGMP regulations. For example, the Office of Management and Budget issued a report that advocated for a uniform set of regulations for all drug products in the interest of promoting efficiency.[28] Similarly, the Office of Technology Assessment issued a report that concluded that the CGMP regulations were too broad and subject to divergent interpretations.[29] Motivated by these observations, FDA proposed substantial rulemaking to update the CGMP regulations.[30]

## 3. Major Revisions to the Regulations (the 1978 CGMP Regulations)

FDA published a proposed set of new CGMP regulations on February 13, 1976 and noted that some of the new requirements in the proposed rules were implicit in the existing regulations.[31] FDA also stated that some of the proposed regulations were "already part of most manufacturers' operations . . . because they are logical extensions of the existing requirements."[32] With respect to production and process controls, however, the agency acknowledged the inherent difficulty in drafting CGMP rules.[33] For example, FDA noted that there were "differing conceptual and statistical approaches" to batch testing and that while "careful attention" was warranted for process controls, articulating a specific standard was challenging and fraught with controversy.[34]

On September 29, 1978, FDA issued final CGMP regulations, which included a substantial preamble containing responses to more than 500 comments.[35] The preamble stated that the CGMP regulations were intended to be general enough to apply to most drug products and also flexible to permit innovation.[36] Generally, the CGMP regulations endeavored to describe results-oriented standards rather than specify exact processes.[37] FDA stated in the preamble that the CGMP regulations were to "describe 'what' is to be accomplished and provide great latitude in 'how' the requirement was achieved."[38] This suggested a results-oriented regulatory system in which industry would be afforded latitude in the implementation of its CGMP requirements. To some extent, this flexible approach seemed contrary to FDA's previously announced goal of promoting unambiguous standards. This flexibility, however, did not apply to areas that had "presented problems in the past."[39] Some CGMPs were, therefore, intended to be applied uniformly in order to achieve compliance.[40]

---

[25] David B. Barr et al., Application of Pharmaceutical CGMPs 10 (1997).

[26] See *id.* at 10-14.

[27] Current Good Manufacturing Practice in Manufacture, Processing, Packing, or Holding, 41 Fed. Reg. 6878, 6878 (Feb. 13, 1976) (to be codified at 21 C.F.R. pts. 210-11).

[28] *Id.*

[29] *Id.*

[30] *Id.*

[31] *Id.*

[32] *Id.* at 6882.

[33] *Id.* at 6882-83.

[34] *Id.*

[35] Current Good Manufacturing Practice in Manufacture, Processing, Packing, or Holding, 43 Fed. Reg. 45,014 (Sept. 29, 1978).

[36] *Id.* at 45,015.

[37] *Id.* at 45,014.

[38] *Id.* at 45,015.

[39] *Id.*

[40] *Id.*

In the preamble, FDA noted that a "current" practice of industry would not necessarily satisfy a particular CGMP requirement.[41] The agency also stated that the purpose of the CGMP authority under the FDCA was to ensure that "drug product quality would not fall below that which was feasible and available under contemporary technology."[42] Uncharacteristically, FDA stated that a "good" practice was essentially one where the benefits of improved drug quality exceeded the cost of the practice.[43] This would almost suggest a cost-benefit analysis element in determining an appropriate practice. The discussion of "current" and "good" implied that FDA would permit variability in production methodology, but would require rigid and increasing quality objectives based on technological innovation in the absence of an explicit CGMP regulation.

A significant portion of the preamble was devoted to discussion of FDA's statutory authority to issue binding CGMP regulations.[44] One should note that the landscape of administrative law was quite different during this time before *Chevron* and the expansion of federal agency authority. In fact, in the preamble, FDA engaged in an extensive discussion on the specific judicial interpretations that granted FDA authority to issue substantive rules under the FDCA.[45] Commenters also challenged the scope of FDA's authority on the basis of the legislative history that stated CGMP regulations were meant to be interpretive and subject to formal rulemaking procedures under section 701(e) of the FDCA.[46]

FDA engaged in a detailed analysis of the FDCA's legislative history and drew several conclusions. First, Congress intended the CGMP statutory authority to enable FDA to issue binding regulations on manufacturers.[47] Second, the regulations could be issued by informal rulemaking.[48] Finally, the agency had broad authority over CGMPs for drug products.[49]

FDA also emphasized in the preamble its belief that a drug product could be adulterated by a failure to comply with CGMP standards generally even if there was no specific CGMP violation related to that product.[50] Indeed, the agency underscored this point saying that "the existence of CGMP regulations does not create violations of the law where they did not previously occur."[51] This would suggest that FDA, despite its substantial CGMP rulemaking, was retaining significant discretion in its enforcement of drug product manufacturing. Under FDA's 1978 interpretation, the CGMP standard under the FDCA could be violated even in the absence of a pertinent regulation.

---

[41] Current Good Manufacturing Practice in Manufacture, Processing, Packing, and Holding, 43 Fed. Reg. at 45,019 (Sept. 29, 1978) (to be codified at 21 C.F.R. pts. 210-11).

[42] *Id.* at 45,020.

[43] *Id.* at 45,025.

[44] *Id.* at 45,020.

[45] *Id.* at 45,023.

[46] *Id.* at 45,020-22.

[47] *Id.*

[48] *Id.* at 45,022-23.

[49] *Id.* at 45,024.

[50] *Id.* at 45,025.

[51] *Id.*

## 4. 1996 CGMP Proposed Revisions

In 1996 FDA again proposed substantial revisions to the CGMP rules. The agency noted that "[r]apid changes in technology . . . created situations not anticipated when the CGMP regulations were . . . last revised."[52] In addition, FDA noted that there was a persistent lack of understanding of certain CGMPs among some drug manufacturers and, as such, greater clarification of the CGMP rules was required.[53] The agency was especially concerned about deficiencies it observed regarding validation.[54] The proposed rules would have covered process validation, methods validation, contamination, testing, and responsibilities of a firm's quality control unit.[55]

In 2002, FDA announced a new initiative to improve its regulatory approach toward pharmaceutical manufacturing.[56] Under this new initiative, "Pharmaceutical CGMPs for the 21st Century," FDA re-evaluated its 1996 proposed rules in light of new developments in product manufacturing standards, such as the quality systems approach and the hazard analysis and critical control points system.[57] The agency ultimately decided to withdraw the 1996 proposed rule and stated it would proceed with CGMP regulations in an "incremental fashion."[58]

## 5. 2007-2008 CGMP Revisions

In conjunction with the withdrawal of its 1996 proposed rule, FDA published a direct final rule that sought to employ the agency's new incremental approach to CGMP regulation.[59] The new rule covered plumbing, aseptic processing, asbestos filters, and process verification.[60] As part of FDA's new initiative the agency stated its intent to modernize and harmonize CGMP standards with international GMP practices.[61] For example, FDA proposed to modify a plumbing requirement standard that required water used in pharmaceutical manufacturing to meet U.S. Environmental Protection Agency (EPA) standards.[62] The agency proposed the new standard, stating that pharmaceutical water must be "safe for human consumption," and that this could be demonstrated by compliance with EPA, European Union, or Japanese potable water standards.[63]

During this rulemaking, FDA received a significant number of negative comments and consequently "withdrew" the direct final rule and addressed the comments in a final rule published in 2008.[64] In the final rule, the agency pared down some of the originally proposed regulations, eliminating the new plumbing rule, and reiterating its policy of incrementally revising its CGMP regulations.[65]

---

52 Current Good Manufacturing Practice; Proposed Amendment of Certain Requirements for Finished Pharmaceuticals, 61 Fed. Reg. 20,104, 20,105 (May 3, 1996).

53 *Id.*

54 *Id.*

55 *Id.* at 20,105-07.

56 Amendment to Current Good Manufacturing Practice; Amendment of Certain Requirements for Finished Pharmaceuticals, 72 Fed. Reg. 68,111, 68,112 (Dec. 4, 2007).

57 *Id.*

58 *Id.*

59 Amendment to Current Good Manufacturing Practice Regulations for Finished Pharmaceuticals, 72 Fed. Reg. 68,064 (Dec. 4, 2007). FDA believed additionally that its CGMP International Harmonization efforts would be better served if it proceeded piecemeal with respect to CGMP rulemaking. *Id.*

60 *Id.* at 68,064-65.

61 *Id.* at 68,065.

62 *Id.*

63 *Id.*

64 Amendments to the Current Good Manufacturing Practice Regulations for Finished Pharmaceuticals, 73 Fed. Reg. 51,919, 51,920 (Sept. 8, 2008) (to be codified at 21 C.F.R. pts. 210-211).

65 *Id.* at 51,919-21.

## 6. Expansion of Statutory Authority under FDASIA

In 2008, FDA began receiving reports of serious adverse reactions, some of which were fatal, in dialysis patients. After an extensive investigation, FDA determined that batches of heparin API manufactured in China were contaminated with oversulfated chondroitin sulfate (OSCS). Moreover, the problem was not limited to crude heparin in the United States. At least 10 other countries were affected. The agency concluded that the contamination of crude heparin was intentional, in order to reduce costs of production. The heparin crisis highlighted other limitations on FDA's ability to ensure product quality.[66] On July 9, 2012, Congress granted FDA the additional statutory authority that FDA sought by including in the Food and Drug Administration Safety and Innovation Act (FDASIA)[67] several provisions related to the drug supply chain.

Section 705 of FDASIA replaced the previous biennial inspection frequency for registered drug firms with a risk-based inspection program.[68] The risk-based model utilizes inspection criteria to prioritize facility inspections. The risk-based criteria include:

- The compliance history of the drug firm;
- The record, history, and nature of recalls linked to the drug firm;
- The inherent risk posed by the drug manufactured, prepared, propagated, compounded, or processed at the drug firm;
- The inspection frequency and history of the drug firm, including whether the firm has been inspected within the last four years;
- Whether the drug firm has been inspected by a foreign government; and
- Any other criteria deemed necessary and appropriate for purposes of allocating inspection resources.[69]

Section 706 amended section 704(a) of the FDCA by allowing FDA to request records outside of an inspection context.[70] Sections 705 and 706 operate in concert, giving FDA the ability to use information collected under its Section 706 authority to determine the risk profile of a facility for purposes of establishment inspection under the Section 705 risk-based model.

FDASIA strengthened FDA's authority regarding inspection refusals so that delaying, denying, limiting, or refusing to permit inspection causes a drug to be adulterated. 21 U.S.C. § 351(j) (a drug is adulterated if "it has been manufactured, processed, packed, or held in any factory, warehouse, or establishment" where the owner has refused to permit entry or inspection). While the "refusal to permit entry or inspection as authorized by section 704" has long been a prohibited act,[71] under Section 707, a refusal now causes all drug products in that establishment to be adulterated.[72]

---

66 Heparin: A Wake-Up Call on Risks to the U.S. Drug Supply, Issue Brief, *Pew Charitable Trust* (May 16, 2012), http://www.pewtrusts.org/en/research-and-analysis/issue-briefs/2012/05/16/heparin-a-wakeup-call-on-risks-to-the-us-drug-supply.

67 The Food and Drug Administration Safety and Innovation Act, Pub. L. No. 112-144, 126 Stat. 993. FDASIA also authorized the collection of FDA user fees for innovator drugs, medical devices, generic drugs and biosimilar biologics. *Id.*

68 The Food and Drug Administration Safety and Innovation Act (FDASIA), Pub. L. No. 112-144, 126 Stat. 993 (2012).

69 *Id.*

70 Food and Drug Administration Safety and Innovation Act (FDASIA), Pub. L. No. 112-144, § 706, 126 Stat. 993, 1067-68 (2012).

71 21 U.S.C. 331(f).

72 Congress required the agency to elaborate on circumstances that would fall within the scope of 21 U.S.C. § 351(j) by issuing guidance within a year of the enactment of FDASIA. See Food & Drug Admin., Guidance for Industry: Circumstances that Constitute

In addition, FDASIA granted FDA explicit authority under Section 710 to engage with foreign counterparts with respect to drug inspections and records exchange.[73] With respect to exchanging records, the legislative history revealed that the provision was designed to help FDA leverage its resources through cooperation with trusted foreign governments.[74] Under Section 710, FDA has the authority to provide information to a foreign government that has been certified as having the ability to protect trade secret information. Under a written agreement with the foreign government, FDA may disclose information related to an establishment inspection for civil regulatory purposes. FDA is also authorized to disclose information to a foreign government for purposes of an investigation, if FDA has reason to believe that a drug could cause serious injury or death.

Section 711 of FDASIA amended section 501 of the FDCA by adding the following language:

> For purposes of paragraph [501](a)(2)(B), the term "current good manufacturing practice" includes the implementation of oversight and controls over the manufacture of drugs to ensure quality, including managing the risk of and establishing the safety of raw materials, materials used in the manufacturing of drugs, and finished drug products.[75]

Section 711 gives FDA the authority to require finished product manufacturers to establish oversight and controls over their suppliers, including contract manufacturing organizations (CMOs). Moreover, if a finished product manufacturer fails to establish oversight and controls related to raw materials, components and contract manufactured finished products, its distributed products are in violation of CGMPs and deemed to be adulterated so that the introduction of those products in interstate commerce is a prohibited act.

The legislative history demonstrates that Congress intended to expand significantly FDA's authority with respect to supply chain management. A House Committee Report on a parallel bill,[76] which contained the same language as section 711, stated that the new statutory text "would incorporate quality risk management oversight and controls into FDA's good manufacturing practices (GMP) authority."[77] The Senate record is even more explicit, and acknowledges FDA's pointed request for additional authority to address the challenges of increasing globalization.[78] Senator Harkin, who sponsored FDASIA, stated that the bill "modernizes FDA's goal of drug supply chain authority" and because "80 percent of all of the ingredients that go into our drugs in this country come from outside this country . . . now we can insist on good manufacturing practices."[79]

### CGMP Regulation Overview

The CGMP regulations are set forth at 21 C.F.R. Parts 210 and 211. Part 210 contains three provisions covering the status, applicability, and definitions pertinent to the CGMP rules. The "status" language in 21

Delaying, Denying, Limiting, or Refusing a Drug Inspection (2014), https://www.fda.gov/downloads/RegulatoryInformation/Guidances/UCM360484.pdf.

[73] Food and Drug Administration Safety and Innovation Act (FDASIA), Pub. L. No. 112-144, §§ 710, 712, 126 Stat. 993, 1070-72 (2012). Section 710 authorizes FDA to disclose certain information to foreign governments, and Section 712 allows recognition of foreign government inspections.

[74] H.R. Rep. No. 112-495, at 33 (2012).

[75] 21 U.S.C. § 351.

[76] Food and Drug Administration Reform Act of 2012, H.R. 5651, 112th Cong. (2012).

[77] H.R. Rep. No. 112-495, at 31 (2012).

[78] See Securing the Pharmaceutical Supply Chain: Hearing Before the S. Comm. on Health, Education, Labor and Pensions, 112th Cong. (2011).

[79] 158 Cong. Rec. S3265 (daily ed. May 17, 2012) (statement of Sen. Thomas Harken); 158 Cong. Rec. S3503 (daily ed. May 23, 2012) (statement of Sen. Thomas Harkin).

C.F.R. § 210.1 states that the codified CGMP regulations are the "minimum current good manufacturing practice" for drug products.[80] In addition, this provision reiterates the provisions of section 501(a)(2)(B) of the FDCA that CGMP violations result in the adulteration of a drug product.[81] Lastly, this regulation states that human cells, tissues, cellular and tissue based products (HCT/Ps) that are drugs are subject to CGMPs in addition to current good tissue practice.[82]

The applicability regulation states that CGMPs are applicable to any drug or drug combination product, and in the event of conflicting obligations for combination products, the more specifically applicable regulation prevails.[83] In addition, this regulation specifies that individuals engaged in some covered operations are only required to comply with CGMPs relevant to their activities.[84] Finally, this provision covers applicability of CGMPs for investigational drugs.[85]

The last section in Part 210 provides various definitions applicable to the CGMP rules.[86]

Part 211 is divided into 11 subparts: Subpart A – General Provisions, Subpart B – Organization and Personnel, Subpart C – Buildings and Facilities, Subpart D – Equipment, Subpart E – Control of Components and Drug Product Containers and Closures, Subpart F – Production and Process Controls, Subpart G – Packaging and Labeling Control, Subpart H – Holding and Distribution, Subpart I – Laboratory Controls, Subpart J – Records and Reports and Subpart K – Returned and Salvaged Drug Products. A discussion of key provisions is set forth below.

**Subpart A – General Provisions**

In this Subpart, FDA states that the regulations in Part 211 contain the minimum requirements for human and animal drug products, except for positron emission tomography (PET) drugs.[87] The rules for PET drugs are set forth in Part 212 of the regulations.

Drugs that are regulated as biological products for human use are regulated under Parts 210-211 and are also regulated under Parts 600-680. Drugs that are also HCT/Ps are regulated under Parts 210-211 and also under Part 1271. Parts 600-680 and Part 1271 supplement, but do not supersede, Parts 210-211.[88]

**Subpart B – Organization and Personnel**

The regulations require each firm to have a quality unit with the responsibility and authority to approve or reject raw materials, in-process materials, finished drug products, and packaging and labeling.[89] In effect, the quality unit serves as the gatekeeper or "traffic cop" for the organization, ensuring that the firm is, and remains in compliance with CGMPs. CGMP regulations do not dictate or require a particular framework for the structure of the quality unit. Indeed, FDA has emphasized that "[t]he CGMP regulations do not subordinate the

---

80 21 C.F.R. § 210.1(a).
81 *Id.* § 210.1(b).
82 *Id.* § 210.1(c).
83 *Id.* § 210.2(a). This regulation is similar to 21 C.F.R. § 211.1.
84 *Id.* § 210.2(b).
85 *Id.* § 210.2(c). For more information see the CGMPs during clinical development discussion.
86 See *id.* § 210.3.
87 *Id.* § 211.1(a).
88 *Id.* § 211.1(b).
89 *Id.* § 211.22.

quality control unit's authority and responsibility to any other unit. . . . They simply require that the quality control unit have final responsibility for certain actions in the manufacturing process."[90] Thus, there is a need for the quality unit to have independence with regard to final decisions related to the quality of the product.[91]

The regulations also require adequate education, training and experience for all personnel, and require firms to have an adequate number of qualified personnel.[92]

**Subpart C – Buildings and Facilities**

The regulations require facilities to "be of suitable size, construction and location to facilitate cleaning, maintenance and proper operations."[93] This Subpart describes, at a high level, the requirements for preventing contamination or product mix-ups throughout the manufacturing process, beginning with the receipt of incoming materials, storage of components and in-process materials, manufacturing, packaging and labeling of products, and quarantine of components, in-process materials and finished product. This Subpart contains requirements for air filtration, environmental monitoring and stringent cleaning. Importantly, 21 C.F.R. § 211.42(d) requires penicillin products to be manufactured in facilities separate from facilities used to manufacture other drug products because penicillin can trigger a severe allergic immune response in some individuals.[94]

Subpart C establishes minimum requirements for lighting, ventilation, plumbing, sewage, washing and toilet facilities, sanitation and maintenance.[95]

**Subpart D – Equipment**

This Subpart requires firms to utilize equipment that is "of appropriate design, adequate size, and suitably located to facilitate operations for its use and for its cleaning and maintenance."[96] Moreover, equipment surfaces that come into contact with components, in-process materials, or finished products cannot have an unintended effect on the characteristics of the product.[97] Substances such as coolants or lubricants used in operating the equipment are not allowed to come into contact with components, in-process materials, or the finished product.[98]

**Subpart E – Control of Components and Drug Product Containers and Closures**

This Subpart requires firms to establish written procedures for the receipt, storage, testing, including approval or rejection, use, and retesting of all incoming raw materials, active ingredients, excipients (coating, binder, flavor, preservatives), containers, and closures.[99]

---

[90] Current Good Manufacturing Practice in Manufacture, Processing, Packing, and Holding, 43 Fed. Reg. 45,014, 45,033 (Sept. 29, 1978) (to be codified at 21 C.F.R. pts. 210-11) (emphasis added).

[91] Note that 21 C.F.R. § 211.22(a) also holds the quality unit responsible for outsourced operations. It is responsible for releasing or rejecting products that are manufactured or packaged by contractors.

[92] 21 C.F.R. § 211.25.

[93] *Id.* § 211.42(a).

[94] FDA requires separate facilities for non-penicillin beta-lactam products for the same reason. See Food & Drug Admin., Guidance for Industry: Non-Penicillin Beta-Lactam Risk Assessment: A CGMP Framework (2013), http://www.fda.gov/ downloads/Drugs/Guidances/UCM246958.pdf.

[95] 21 C.F.R. § 211.44-58.

[96] *Id.* § 211.63.

[97] *Id.* § 211.65(a).

[98] *Id.* § 211.65(b).

[99] *Id.* § 211.80-87.

**Subpart F – Production and Process Control**

Subpart F requires the firm's quality control unit to establish written production and process control procedures "designed to assure that the drug products have the identity, strength, quality and purity they purport or are represented to possess."[100] The regulations require production and process control procedures to be followed during the entire manufacturing process, and completion of process steps must be documented at the time of performance.[101] Any deviation from these procedures must be recorded and justified.[102] This Subpart also requires firms to have written procedures requiring testing of in-process materials and drug products to ensure uniformity and integrity of each batch.

**Subpart G – Packaging and Labeling Control**

This Subpart requires the establishment of written procedures to assure that the correct labels, labeling and packaging materials are used for the drug product.[103] Each drug product must have a lot or control number.[104] These requirements are designed to prevent label mix-ups. This Subpart also contains requirements related to product expiration dating. The expiration date is determined based on stability testing, which evaluates the drug product's ability to remain stable (i.e., not degrade or become less potent) over time. The product expiration date assures that the drug product meets applicable standards of identity, strength, quality and purity at the time of use.[105]

**Subpart H – Holding and Distribution**

This Subpart requires firms to have written procedures for the quarantine of products prior to release by the control quality unit,[106] and for storage of products under appropriate temperature, humidity and lighting conditions.[107] The Subpart also requires firms to have written procedures for distribution of drug products (oldest product to be distributed first),[108] and for a system that allows each lot of a product to be readily identified in the event a recall is required.[109]

**Subpart I – Laboratory Controls**

This Subpart requires "the appropriate organizational unit," e.g., the Microbiology Laboratory or the Analytical Chemistry Laboratory, to draft and submit all written laboratory controls to the quality unit for approval. Section 211.160(b) states that such controls "shall include the establishment of scientifically sound and appropriate specifications, standards, sampling plans, and test procedures designed to assure that components, drug product containers, closures, in-process materials, labeling and drug products conform to appropriate standards of identity, strength, quality and purity." The Subpart includes requirements for testing and release for distribution, stability testing, maintaining reserve samples, and testing for penicillin contamination. With regard to stability testing, the regulations state that the testing program must be written, and requires that stability test results be used to determine storage conditions and expiration dates.[110]

---

[100] *Id.* § 211.100(a). The failure to establish adequate procedures for production and process controls can have a significant adverse impact on product quality and patient safety. For example, such controls are essential to prevent glass or particulate contamination in sterile injectable products.

[101] This is widely described as "concurrent documentation."

[102] 21 C.F.R. § 211.100(b).

[103] *Id.* § 211.122.

[104] *Id.* § 211.130.

[105] *Id.* § 211.137(a).

[106] *Id.* § 211.142(a).

[107] *Id.* § 211.142(b).

[108] *Id.* § 211.150(a).

[109] *Id.* § 211.150(b).

[110] *Id.* § 211.166.

**Subpart J – Records and Reports**

This Subpart establishes requirements for production, control, and distribution records that are required to comply with CGMP regulations. These records include equipment cleaning and use logs, component, container closure and labeling records, master production and control records, batch production and control records, laboratory records, distribution records and complaint files.

The quality unit's responsibility for reviewing all production records encompasses product, control, packaging, and labeling records. As stated in the regulations:

> [a]ll drug product production and control records, including those for packaging and labeling, shall be reviewed and approved by the quality control unit to determine compliance with all established, approved written procedures before a batch is released or distributed. Any unexplained discrepancy (including a percentage of theoretical yield exceeding the maximum or minimum percentages established in master production and control records) or the failure of a batch or any of its components to meet any of its specifications shall be thoroughly investigated, whether or not the batch has already been distributed. The investigation shall extend to other batches of the same drug product and other drug products that may have been associated with the specific failure or discrepancy. A written record of the investigation shall be made and shall include the conclusions and followup.[111]

The regulations require a review of production records before release or distribution of product batches, and, in the case a discrepancy is found, a full investigation of all batches associated with the failure. As part of the investigation, the quality unit is required to develop a written investigation report. The regulations also require that "[w]here an investigation under § 211.192 is not conducted the written record shall include the reason that an investigation was found not to be necessary and the name of the responsible person making such a determination."[112] Thus, FDA seeks to hold the individuals within the quality unit making the determination that an investigation is not necessary accountable for their decision by requiring that their names be identified.

As part of the obligations under 21 C.F.R. §§ 211.192 and 211.198 that discrepancies and batch or component failures be investigated and recorded, FDA, through its regulations, requires that procedures be in place to put responsible officials on notice of any investigations conducted or found unnecessary. The regulations state:

> [p]rocedures shall be established to assure that the responsible officials of the firm, if they are not personally involved in or immediately aware of such actions, are notified in writing of any investigations conducted under 211.198, 211.204, or 211.208 of these regulations, any recalls, reports of inspectional observations issued by the Food and Drug Administration, or any regulatory actions relating to good manufacturing practices brought by the Food and Drug Administration.[113]

Records and reports are key indicators of a firm's compliance with CGMPs. As discussed in Chapter II, the review of manufacturing records is a significant part of any CGMP establishment inspection.

**Subpart K – Returned and Salvaged Drug Products**

This Subpart requires firms to establish controls for storage, testing, and reprocessing of returned products. Under 21 C.F.R. § 211.204, if storage or shipping conditions call into question the "safety, identity, strength,

---

[111] *Id.* § 211.192.

[112] *Id.* § 211.198(b)(3) (emphasis added).

[113] *Id.* § 211.180(f) (emphasis added).

quality or purity" of the drug product, the regulations require the product to be destroyed unless testing, examination of the product or other investigation confirms that the product meets all quality attributes. This section establishes provisions for reprocessing, recordkeeping and investigation of associated batches if the return was due to a problem that implicates other batches.[114] Section 211.208 establishes requirements for salvaging products that have been subjected to improper storage conditions as the result of a disaster, fire, accident, or equipment failure.

## 7. CGMP Case Law

In 1968, the Eastern District of New York, in *United States v. Bel-Mar Laboratories, Inc.*, 284 F. Supp. 875-83 (E.D.N.Y 1968) held that the CGMP standards were sufficiently definite and clear so as to be understood by the industry and necessitate compliance.[115] In *Bel-Mar*, the defendant manufacturer was charged with unlawfully introducing adulterated injectable drugs into interstate commerce in violation of the FDCA.[116] In seeking to dismiss the charges, the manufacturer argued that section 351(a)(2)(B) of the FDCA, which requires conformity with CGMPs, was "devoid of any definite, certain or ascertainable standards with which one could determine whether the methods being used in, or the facilities or controls being used for the manufacture, processing, packing or holding of a particular drug conform to, or are being operated or administered in conformity with 'current good manufacturing practice.'"[117] The court disagreed and found that "the phrase 'current good manufacturing practice' is not strange to those in the trade to whom the subject section is directed"[118] and that FDA had promulgated guidance and interpretive regulations sufficient to set minimum requirements for the industry and that these were even "welcomed by some in the industry because they serve[d] to focus attention on the problem areas, and to guide in the evaluation of a product's integrity."[119] Thus, CGMPs and their guidance were found by the court to be sufficiently definite to overcome the manufacturer's argument that these were unconstitutionally vague.[120]

Three years later, the court in *United States v. Lit Drug Co.*, 333 F. Supp. 990-97 (D.N.J. 1971) further defined the breadth of the CGMP requirements of the FDCA by holding that a drug could be deemed adulterated under the FDCA even where it was "pharmaceutically perfect in content" if the drug was "processed, packed or held under unsanitary conditions, or if any manufacturing, packing or holding method does not conform to current good manufacturing practice."[121] In *Lit Drug*, the court upheld an injunction that prohibited the continued interstate sale of the drug product until FDA could complete a thorough inspection of "all equipment, materials, labels, containers and procedures" including an examination and assay of all drugs held at the manufacturer's plant.[122] Accordingly, post-*Lit Drug*, it was clear to the industry that the manner in which a drug was produced was as important as the composition of the drug itself, and thus, violative manufacturing

[114] As a practical matter, if storage or shipping conditions call into question the quality of drug products, most firms choose to destroy the product rather than re-release based on retesting or investigation because of the risk that the product might fail stability testing later in the expiry period.

[115] 284 F. Supp. 875, 883 (E.D.N.Y. 1968).

[116] *Id.* at 878.

[117] *Id.* at 879.

[118] *Id.* at 883.

[119] *Id.*

[120] *Id.* at 882-83.

[121] 333 F. Supp. 990, 998 (D.N.J. 1971).

[122] *Id.* at 997-1000.

practices could serve as the basis for an enforcement action even if the pharmaceutical integrity of the drug was not ultimately affected.

Soon after the 1978 CGMP regulations were published, the pharmaceutical industry took FDA to court to challenge the validity of its rulemaking. In *National Association of Pharmaceutical Manufacturers* [*NAPM I*] industry sought declaratory relief to find that the CGMP regulations were only interpretive and thus only constituted prima facie evidence of adulteration.[123] The plaintiffs in the case based their position on legislative history that suggested FDA was limited to issuing interpretive CGMPs.[124] The court, however, found the plaintiff's discussion of the legislative history was taken out of context, and therefore engaged in a thorough review of the Kefauver-Harris Amendments, which largely mirrored FDA's previous analysis under the CGMP rulemaking.[125] In addition, the court found the reasoning unpersuasive in the two cases[126] the plaintiffs offered, which suggested CGMPs were only interpretive.[127] The court ultimately held that the CGMPs were authorized to be issued as binding regulations.[128]

Industry would, however, challenge the CGMP regulations again. In another suit, *National Association of Pharmaceutical Manufacturers* [*NAPM II*], industry attacked the CGMPs on a variety of grounds, such as: CGMPs did not apply to new drugs; adulteration could not necessarily be attributable to CGMP violations; FDA incorrectly interpreted what CGMPs meant; CGMPs could not be a static standard as they evolve over time; CGMPs were too vague; FDA violated procedural requirements in its rulemaking; there was an insufficient administrative record; and FDA was arbitrary and capricious in its rulemaking.[129] The court in *NAMP II* dismissed many of the complaints on grounds of res judicata.[130] Specifically, the court found that claim preclusion applied to the question of whether CGMPs were binding and substantive regulations, an issue the court found already decided by *NAMP I*.[131]

The court in *NAMP II* then proceeded to analyze the plaintiff's other claims.[132] Each of the non-precluded claims was examined with the court undertaking an analysis of the Kefauver-Harris Amendments and FDA's actions in the CGMP rulemaking.[133] In the analysis, the court generally found that FDA had appropriately engaged in rulemaking for CGMPs and the regulations were not invalid under any of the plaintiff's contentions.[134] The court ultimately granted the government's motion for summary judgment on the grounds of either res judicata or on the merits.[135]

In 1988, another federal court had the opportunity to consider the validity of FDA's CGMP regulations. On this occasion, the dispute involved the withdrawal of a company's applications based on CGMP

[123] Nat'l Ass'n of Pharmaceutical Mfrs. v. Food and Drug Admin., 487 F. Supp. 412, 414 (S.D.N.Y. 1980).

[124] *Id.* at 414-15.

[125] *Id.*

[126] *Id.* at 415 (citing United States v. An Article of Drug Labeled White Quadrisect, 484 F. 2d 748 (7th Cir. 1973) and United States v. Bel-Mar Labs. Inc., 284 F. Supp. 875 (E.D.N.Y. 1968)).

[127] *Id.* at 415.

[128] *Id.*

[129] Nat'l Ass'n of Pharmaceutical Mfrs. v. Dep't of Health and Human Servs., 586, F. Supp. 740, 744-46 (S.D.N.Y. 1984).

[130] *Id.* at 746.

[131] *Id.*

[132] *Id.* at 747.

[133] See *id.* at 747-60.

[134] *Id.*

[135] Nat'l Ass'n of Pharmaceutical Mfrs., 586 F. Supp. at 763.

violations.[136] As discussed in Chapter III, section 355(e) of the FDCA governs the withdrawal of NDA approval, and requires that FDA provide applicants with due notice and opportunity to be heard prior to withdrawal. However, a hearing may not be necessary where there is no "genuine and substantial issue of fact" and summary withdrawal of approvals is appropriate.[137] For example, where a company has been the subject of numerous inspections and been made aware that CGMP violations exist, its failure to correct within a reasonable time may be sufficient to permit the summary withdrawal of its applications.

In *Copanos and Sons, Inc. v. FDA*, the D.C. Circuit concluded that because the defendants received adequate notice of the basis for FDA's withdrawal of its application for failure to comply with CGMPs, withdrawal pursuant to summary judgment, i.e., without a hearing, was appropriate and not subject to review.[138]

The defendants, John D. Copanos & Sons, Inc. and an affiliated entity, Kanasco, Ltd. (collectively referred to here as Copanos), holders of NDAs and New Animal Drug Applications (NADAs), were the subject of numerous CGMP inspections by FDA over the course of four years.[139] The first inspection of Copanos's oral dosage manufacturing facility revealed CGMP violations including falsified manufacturing records, problems with the calibration of laboratory equipment, insanitary manufacturing conditions, and production of sub-potent and impure animal drugs.[140] As a result of these findings, FDA issued a Form FDA-483 and sought to enjoin Copanos from continuing to manufacture and ship drug products.[141] FDA dismissed the complaint with prejudice after Copanos voluntarily agreed to suspend manufacturing or shipment of the drug products without prior written approval from FDA.[142] Copanos also submitted a response to the Form FDA-483 acknowledging the deficiencies and promising to implement corrective action.[143]

Approximately five months later, FDA again inspected the facility, found the same CGMP manufacturing violations were not properly or fully addressed, and issued a second Form FDA-483.[144] Though Copanos responded to the Form FDA-483 and again promised to implement corrective actions, FDA, not persuaded by the response, denied Copanos's request to resume manufacturing.[145] The pattern of inspections and failure to correct to FDA's satisfaction continued, during which time it was determined that Copanos manufactured and shipped more than one million vials of injectable drug products in violation of the voluntary agreement to suspend distribution without prior written approval,[146] and withheld documents that would have alerted FDA of such shipments.[147]

---

[136] Copanos and Sons, Inc. v. Food and Drug Admin., 854 F.2d 510 (D.C. Cir. 1988).

[137] See 21 C.F.R. § 314.200(g).

[138] Copanos and Sons, Inc., 854 F.2d at 514.

[139] *Id.* at 515-518.

[140] *Id.* at 515.

[141] *Id.*

[142] *Id.*

[143] *Id.*

[144] *Id.*

[145] *Id.* Copanos specifically requested authorization to resume the manufacturing, though not shipping, of its animal drugs products arguing that these drugs were manufactured in compliance with CGMP. FDA, however, was not satisfied with the media fill performed and other laboratory controls noting that the "firm does not have an adequate number of individuals with sufficient training and experience to perform the necessary quality control, quality assurance, and laboratory activities." *Id.*

[146] *Id.* at 516. Pursuant to the prior Agreement, Copanos has agreed to manufacture only one batch of its injectable drug products for the limited purposes of testing its manufacturing process.

[147] *Id.* at 516.

Soon after this discovery, FDA filed a second complaint for injunctive relief and proceeded to conduct a search of the manufacturing company pursuant to a criminal search warrant, which identified additional CGMP violations.[148] By consent, the district court entered an interim order prohibiting the company from the manufacturing and shipping of injectable drug products in the absence of prior authorization, and Copanos was later ordered to show cause why this injunction should not be made permanent.[149] Rather than respond to the show cause and litigate the permanent injunction action, Copanos agreed to enter into a consent decree, which prohibited the manufacture and distribution of drugs while the CGMP violations persisted and required specific action.[150]

Only weeks after the entry of the consent decree, an FDA inspection again revealed that the same CGMP violations persisted, ultimately leading FDA to seek a court order requiring that the company cease all manufacturing of the drug.[151] Copanos again promised to correct the deficiencies and FDA authorized the resumption of manufacturing and shipping.[152] Less than a year later, FDA investigators found the same CGMP issues during an inspection.[153]

The government sought enforcement of the consent decree, which eventually, by consent, became permanent.[154] In an attempt to secure authorization to reopen the facility, the company submitted validation data.[155] An inspection conducted a month later, however, revealed both new and persistent CGMP violations.[156] A month later, FDA proposed to withdraw the company's NDAs and NADAs for injectable products via a notice published in the Federal Register.[157]

Pursuant to the notice, the company requested a hearing and moved for summary judgment on the basis that it had not been provided proper written notice "specifying the matters complained of" and "a reasonable time after receipt of written notice" in which to cure any deficiencies.[158] The request was later denied, and FDA ordered that the approval of the NDAs and NADAs for sterile injectable products be withdrawn from the market.[159] Copanos argued that violations were not specifically detailed or, alternatively that FDA's authorization to continue to manufacture established compliance with CGMPs. The court found, however, that because "the record demonstrate[d] that in several respects [Copanos's] methods, facilities and controls were 'inadequate and were not made adequate within a reasonable time after receipt of written notice specifying the matter complained of'.... FDA was ... entitled to withdraw [Copanos's] NDAs and NADAs for injectable drugs without a hearing."[160] The court held so despite acknowledging that "[t]he CGMP regulations on which the FDA based its withdrawal order in this case are 'imprecise' in various places,"[161] noting that any factual disputes created by the imprecise language were not material.[162]

---

[148] *Id.*
[149] *Id.*
[150] *Id.*
[151] *Id.* at 516-517.
[152] *Id.* at 517
[153] *Id.*
[154] *Id.*
[155] *Id.*
[156] *Id.*
[157] *Id.*
[158] *Id.* at 518.
[159] *Id.*
[160] *Id.* at 526.
[161] *Id.* at 522.
[162] *Id.*

The court also rejected Copanos's alternative argument that deficiencies observed prior to the entry of the consent decree could not be the basis for liability.[163] FDA, Copanos argued, essentially "certified" its compliance with CGMP regulations at the time the consent decree was entered and therefore it was no longer on notice of any violations.[164] In rejecting Copanos's argument, the court stated that adopting such an interpretation of the regulations and statute "would allow manufacturers that chronically violate CGMP to avoid the withdrawal of the approvals by momentarily 'correcting' the deficiencies each time they are noted by the FDA."[165] The court further noted, "requiring current non-compliance would substantially diminish FDA's ability to move against manufacturers who repeatedly violate CGMP but pull themselves temporarily into compliance each time the FDA begins the process of withdrawing their approvals."[166] Ultimately, given Copanos's historical record of non-compliance, the court held that no material disputes of fact existed and therefore, FDA was entitled to withdraw the company's NDAs and NADAs.[167]

*Barr Laboratories*

On February 5, 1993, the United States District Court for the District of New Jersey issued its opinion in *United States v. Barr Laboratories, Inc.*, widely considered the landmark case on the CGMP requirements.[168] In its summary of the dispute, the court acknowledged the conflicting but legitimate interests of the parties by stating:

> Each day with confidence and hope millions of people in the United States and other countries reach for pills, powders, capsules and syrups to relieve or prevent an infinite number of physical and mental ailments. The weighty task of ensuring the integrity of these products, frequently unquestioned by most consumers, falls to the Food and Drug Administration, which monitors the practices of the drug industry through a system of approvals and investigations. Built into this maze of often ambiguous rules, however, is the recognition that drug manufacturers are businesses, which must follow efficient as well as effective procedures.[169]

Acknowledging the contentious nature of the litigation, the court stated further that "[t]he current conflict surrounding these rules is best characterized as a confrontation between a humorless warden and his uncooperative prisoner."[170] The *Barr* trial consumed 15 days between August 17 and October 12, 1992. In response to the government's allegations of CGMP violations, Barr argued that the requirements that FDA sought to impose were neither required nor contemplated by the FDCA or the CGMP regulations.[171] Barr also argued that its laboratory and manufacturing practices were consistent with industry standards.[172] The evidence presented consisted of more than 2,300 pages of testimony, roughly 400 documents and numerous written declarations.[173] The court was left with the task of sifting through all the evidence and determining the appropriate standards to apply.

---

[163] *Id.* at 525.
[164] *Id.*
[165] *Id.* at 525.
[166] *Id.*
[167] *Id.* at 526.
[168] United States v. Barr Labs., Inc., 812 F. Supp. 458 (D.N.J. 1993).
[169] *Id.* at 464.
[170] *Id.*
[171] Barr Findings of Fact and Conclusions of Law (Proposed) ¶¶ 42-48, at 97-99, United States v. Barr Labs., Inc., 812 F. Supp. 458 (D.N.J. 1993) (No. 92-1744).
[172] *Id.* ¶¶ 5- 6, at 3-4.
[173] Barr Labs., 812 F. Supp. at 464.

The court recognized that CGMP regulations are "[d]esigned as a quality control measure" and "outline general rules for all aspects of drug manufacture including buildings and facilities, personnel, equipment, drug components and containers, production, packaging and labeling, and record-keeping."[174] Most importantly, the court stated that "failure to comply with CGMP regulations renders any resulting drug product 'adulterated' and the drug product and its producer subject to regulatory action."[175]

At trial, FDA and Barr offered quite different interpretations of what the regulations required. FDA argued that "CGMPs are those manufacturing practices and controls which are currently observed by the better firms in the industry."[176] FDA also argued that to the extent that the regulations are silent on a point, manufacturers are required to use FDA guidelines, as well as scientific literature, to determine what constitutes CGMP.[177] By contrast, Barr argued that the CGMP regulations are "general in nature and subject to considerable interpretation. . . . There are many ways to satisfy the objectives of CGMP, and a firm cannot fairly be said to be in violation of CGMP so long as that firm's approach is rational and not violative of any general principles of CGMP."[178]

The court agreed with Barr that the CGMP regulations set the parameters for compliance, allowing for "conflicting, but plausible" interpretations of what constitutes CGMP.[179] The court agreed with FDA, however, that to the extent the regulations are vague, companies must look to scientific literature or "employ scientific judgment where appropriate."[180]

Perhaps the most contentious issue in the case was the definition of the word "failure." FDA's initial position was that any out-of-specification (OOS) result, even in an early stage of a multistage test procedure, was a "failure" requiring rejection of the entire production batch.[181] Had the court accepted this definition, the U.S. pharmaceutical industry could not have survived. As the trial wore on, FDA retreated from this initial position and agreed, for example, that an OOS result at the first stage of a content uniformity test (prescribed by the U.S. Pharmacopeia (USP) as a two-stage test) or at the first or second stage of a dissolution test (a three-stage USP test) is not a "failure."[182] The government also acknowledged that an OOS result attributable to laboratory error would not be a "failure."[183]

Barr argued that if an OOS result were due to an identified error, it could be invalidated and the test repeated. Even when an OOS result could not be attributed to an identified laboratory error, Barr argued, it could be overcome through statistical analysis or by retesting.[184]

The court concluded that the term "out-of-specification result" referred to "an individual test value that does not meet predetermined specifications."[185] Because an OOS result could be attributed to laboratory error

---

[174] *Id.* at 465 (emphasis added).

[175] *Id.*

[176] Government's Proposed Findings of Fact and Conclusions of Law ¶ 8, at 3, United States v. Barr Labs., Inc., 812 F. Supp. 458 (D.N.J. 1993) (No. 92-1744) (emphasis added).

[177] *Id.*

[178] Barr Findings of Fact and Conclusions of Law (Proposed), *supra* note 166, ¶¶ 5- 6, at 3-4.

[179] Barr Labs., 812 F. Supp. at 465.

[180] *Id.*

[181] Transcript at 132-33, United States v. Barr Labs., Inc., 812 F. Supp. 458 (D.N.J. 1993) (No. 92-1744).

[182] *Id.* at 1044.

[183] *Id.* at 225.

[184] Barr Findings of Fact and Conclusions of Law (Proposed), *supra* note 166, ¶¶ 49, 51, at 22-24.

[185] Barr Labs., 812 F. Supp. at 466, n.2.

or could be discredited through the use of an outlier test or appropriate retesting, the court rejected the government's definition of "failure."[186]

Barr also argued that the use of outlier tests to determine whether a test result is statistically consistent with other test results was "scientifically reasonable." [187]Barr called the court's attention to specific language in the CGMP regulations requiring that "statistical quality control criteria . . . include appropriate acceptance levels and/or appropriate rejection levels."[188] What is "appropriate," Barr argued, was a matter of scientific judgment.[189] FDA challenged Barr's use of statistical outlier tests, calling it "inappropriate and statistically invalid."[190]

Although the court agreed that outlier tests could be used to invalidate OOS results, the court determined that the use of outlier testing should be restricted to microbiological assays.[191] The court's reasoning was that the USP expressly allowed the use of outlier tests for microbiological assays but did not discuss their use for chemical tests.[192] The USP's silence on the latter point, said the court, constituted a prohibition.[193] Thus, on this issue, FDA prevailed in eliminating outlier tests in the chemical area, and Barr prevailed in preserving outlier tests for microbiological assays.

With regard to testing, historically, if an analyst generated an OOS result, Barr's laboratory would order retesting on the same sample by two different chemists [hereinafter Barr's "two out of three" practice]. Prior to 1992, Barr averaged the results obtained on retest and used that average as the result of record. In 1992, Barr began including the initial OOS result in the average for a result of record.[194] FDA argued that whether the initial OOS result was discarded or included in the final result was irrelevant: both practices were "unscientific and unsound."[195]

FDA faulted Barr's retest procedure on two grounds. First, FDA argued that Barr often conducted retests on different samples (allegedly non-homogeneous parts of the batch), and thus, retesting might obscure variability within a batch.[196] FDA's argument was based on its belief that a "sample" was a single 20-tablet grind. If an analyst selected a new group of 20 tablets from the same bottle, FDA argued, that was a "resample."[197]

Barr's position was that the entire bottle brought to the laboratory by Quality Assurance, in which tablets or capsules were collected throughout the manufacturing run, constituted the "sample,"[198] and that any group of 20 tablets from that bottle is part of the same "sample."[199] Barr argued that "resampling" means going to

---

[186] *Id.* at 467.

[187] Barr Findings of Fact and Conclusions of Law (Proposed), *supra* note 166, ¶ 51, at 23.

[188] 21 C.F.R. § 211.165(d) (emphasis added).

[189] Barr Findings of Fact and Conclusions of Law (Proposed), *supra* note 166, ¶ 51, at 23.

[190] Government's Proposed Findings of Fact and Conclusions of Law, *supra* note 171, ¶ 35, at 15-16.

[191] Barr Labs., 812 F. Supp. at 469.

[192] *Id.*

[193] *Id.*

[194] Transcript, *supra* note 176, at 2192-94.

[195] Government's Proposed Findings of Fact and Conclusions of Law, *supra* note 171, ¶ 31, at 12.

[196] *Id.* ¶ 17-27, at 7-10.

[197] Transcript, *supra* note 176, at 1022.

[198] *Id.* at 1533-34.

[199] *Id.*

another bottle, selected from retains, that had not been filled throughout the production run.[200] The court adopted Barr's view:

> A retest is defined as additional testing on the same sample, and thus it necessarily follows an initial test. An analyst performing a retest takes the second aliquot from either: (1) the sample that was the source of the first aliquot ... or (2) the larger sample previously collected for laboratory purposes ... These procedures are equivalent.[201]

FDA's second criticism of Barr's retest procedure was that Barr substituted retesting for an appropriate failure investigation.[202] FDA took the position that most failures are process-related, and therefore retesting only served to mask process-related problems.[203] In FDA's view, retesting would be appropriate only where a failure investigation had identified definitively a specific laboratory (and hence non-process-related) error.[204]

Barr argued that retesting was an important means of investigating an OOS result.[205] "Additional testing represents an important means of ascertaining whether an out-of-specification test result is due to analytical or other error unrelated to the quality of the product .... In testing for potency within a sample, increased testing heightens the likelihood that the true value will be ascertained."[206] With respect to its two-out-of-three approach, Barr produced evidence that other pharmaceutical companies, both "brand-name" and generic, used the same method for retesting.[207]

The court found that "[r]etesting is necessary if a failure investigation indicates that analyst error caused an initial out-of-specification result."[208] Adopting Barr's view, the court held that as part of an investigation "[a] retest is similarly acceptable when review of the analyst's work is inconclusive."[209] In such cases, retesting "substitutes for or supplements the original test results."[210] The court determined that the actual number of retests necessary to confirm or reject an OOS result is a matter of scientific judgment.[211] However, the court stated that firms must do "enough testing" to reach a conclusion.[212] "Such a conclusion cannot be based on 3 of 4 or 5 of 6 passing results, but possibly 7 of 8."[213]

Finally, the court stated that a proper testing procedure must establish in advance a point at which all testing will cease and the results will be evaluated.[214]

---

200 *Id.*

201 United States v. Barr Labs., Inc., 812 F. Supp. 458, 469 (D.N.J. 1993).

202 Transcript, *supra* note 176, at 57.

203 *Id.* at 60, 65-66.

204 Government's Proposed Findings of Fact and Conclusions of Law, *supra* note 171, ¶ 13, at 5-6.

205 Barr Findings of Fact and Conclusions of Law (Proposed), *supra* note 166, ¶ 45, at 20.

206 *Id.*

207 *Id.* ¶ 46, at 21.

208 United States v. Barr Labs., Inc., 812 F. Supp. 458, 469 (D.N.J. 1993).

209 *Id.*

210 *Id.*

211 Barr Findings of Fact and Conclusions of Law (Proposed), *supra* note 166, ¶ 46, at 21.

212 Barr Labs., 812 F. Supp. at 470.

213 *Id.* The court rejected Barr's two-out-of-three retest procedure even though Barr produced evidence that the procedure was used throughout industry, "[t]he problems with Barr's retesting procedure are not erased by the protocol's popularity at other firms. ... . [I]ndustry practice while instructive also is subject to the CGMP-compliance test." *Id.* at 470, n.9.

214 *Id.* at 470.

The court rejected Barr's view that averaging test results is an acceptable industry practice. Averaging, said the court, masks variability among the data.[215] The court was particularly concerned about averaging chemical assay test data where some of the initial values are within specification and others are not.[216] In such instances, the average result might be within specification, and the batch released, even though individual values are out of specification.[217] The court was less concerned about microbiological assays, stating that for microbiological assays, "the USP prefers an average . . . when reaching an ultimate judgment about the quality the product."[218] Thus, Barr lost its averaging argument for chemical testing but won the point for microbiological testing.

FDA criticized both the content and the timeliness of Barr's failure investigations.[219] According to FDA, a full-scale investigation was required for every OOS result.[220] Barr disputed these claims.[221] In Barr's view, the appropriate scope of a particular investigation depended on the type of problem identified.[222] Barr suggested to the court that when an analyst obtained an OOS result, the analyst and the supervisor would review the analyst's calculations and inspect the laboratory instruments used in the procedure as the first step in the laboratory investigation.[223] If the laboratory could identify the cause of the problem at that stage, the laboratory would invalidate the OOS result. If the laboratory could not identify the cause of the problem, the laboratory would order retests by two different analysts.[224]

Barr argued that the two-out-of-three retesting constituted the laboratory investigation, and the test results were the only documentation needed.[225] If the test results were inconsistent, the laboratory would conduct an appropriate outlier test and then average all consistent results.[226] If the average of all consistent results was within specification, the lab considered the investigation complete and released the batch.[227] If the average was OOS or if one or both of the retest analysts obtained an OOS result, the laboratory notified Quality Assurance to open a formal investigation.[228]

The court agreed with Barr that the proper scope of an investigation depended upon the issue presented. The court said that all investigations should be documented fully and conducted within 30 business days. However, the court set certain limits on industry's ability to discredit OOS results by restricting the use of outlier tests and by rejecting Barr's two-out-of-three testing procedure.[229]

---

[215] *Id.* at 471.

[216] *Id.*

[217] *Id.*

[218] *Id.*

[219] Government's Proposed Findings of Fact and Conclusions of Law, *supra* note 171, ¶ 65, at 31.

[220] *Id.* ¶ 59, at 28. The government argued that an adequate failure investigation required a written analysis that included: 1) identification of the problem; 2) a discussion of possible causes; 3) proposed corrective actions; 4) identification of other batches that might have been affected; 5) comments from Quality Control and Manufacturing upon completion of any corrective actions; and 6) signatures of employees responsible for the investigation. Barr Labs., 812 F. Supp. at 467.

[221] Transcript, *supra* note 176, at 1558.

[222] *Id.* at 1524-36.

[223] Barr Findings of Fact and Conclusions of Law (Proposed), *supra* note 166, ¶ 54, at 24-25.

[224] Transcript, *supra* note 176, at 1527.

[225] *Id.* at 1528.

[226] *Id.* at 1529.

[227] *Id.*

[228] Barr Findings of Fact and Conclusions of Law (Proposed), *supra* note 166, ¶ 56, at 25.

[229] United States v. Barr Labs., Inc., 812 F. Supp. 458, 478-79 (D.N.J. 1993).

Another issue at trial related to Barr's validation of analytical methods. An "analytical method" is a laboratory procedure that measures an attribute of a product, component or raw material.[230] "Analytical methods validation" is the process of demonstrating that an analytical method is reliable and capable of performing the analysis for which the method has been designed.[231] The validation process consists of experiments to determine certain parameters, including precision, ruggedness, accuracy, selectivity, and linearity.[232]

Section 211.165(e) of the CGMP regulations states, and stated at the time of the Barr trial, that:

> The accuracy, sensitivity, specificity, and reproducibility of test methods employed by the firm shall be established and documented. Such validation and documentation may be accomplished in accordance with § 211.194(a)(2).

Section 211.194(a)(2) provides that:

> (a) Laboratory records shall include complete data derived from all tests necessary to assure compliance with established specifications and standards, including examinations and assays, as follows:
>
> (2) A statement of each method used in the testing of the sample. The statement shall indicate the location of data that establish that the methods used in the testing of the sample meet proper standards of accuracy and reliability as applied to the product tested. (If the method employed is in the current revision of the United States Pharmacopeia, . . . or in other recognized standard references, or is detailed in an approved new drug application and the referenced method is not modified, a statement indicating the method and reference will suffice.) The suitability of all testing methods used shall be verified under actual conditions of use.

Prior to trial, FDA criticized Barr's analytical method validation studies for a variety of reasons.[233] At trial, however, neither of the government's witnesses claimed to have any expertise in analytical methods or analytical methods validation.[234] Moreover, the only evidence submitted by the government from an analytical chemist was a written declaration that did not criticize any of Barr's methods.[235]

The court found that an analytical method can be deemed validated if: 1) the method was approved as part of a firm's ANDA (or, consequently, for a brand-name company, its NDA); 2) the method is identical to the current USP method[236] or 3) a firm conducts a satisfactory validation study.[237] The court also found that if a firm uses a method not recognized by USP or modifies a USP method, the firm must validate the method.[238]

Because Barr's methods conformed to the USP or the Code of Federal Regulations (CFR), or were approved as part of Barr's ANDAs, the court determined that the methods were entitled to a "presumption of validity."[239] To the extent that Barr and the government disagreed about certain aspects of Barr's validation studies,

---

[230] Transcript, *supra* note 176, at 1747.

[231] *Id.* at 1759.

[232] *Id.* at 1763-65.

[233] Barr Findings of Fact and Conclusions of Law (Proposed), *supra* note 166, ¶ 7, at 4.

[234] *Id.* ¶ 8, at 4.

[235] *Id.*

[236] Or, for antibiotic products, the applicable method set forth in the Code of Federal Regulations (CFR).

[237] United States v. Barr Labs., Inc., 812 F. Supp. 458, 482 (D.N.J. 1993).

[238] *Id.*

[239] *Id.*

the court advised the parties to resolve their differences informally, stating "the court is confident that the parties will be able to resolve any subsequent disagreements without judicial intervention."[240]

The Barr court's interpretation of CGMP regulations was a fundamental departure from the "flexible" approach adopted by FDA in the 1978 CGMP regulations. Stating in the opinion that the lack of clear regulatory standards had resulted in "an industry mired in uncertainty and conflict, guided by vague regulations which produce tugs-of-war of varying intensity,"[241] the court proceeded to establish explicit parameters for drug testing, which included:

- What constitutes a "failure"
- The process for conducting the laboratory investigation of an OOS result
- The process for conducting the formal investigation of confirmed OOS results and other manufacturing problems
- Appropriate use of outlier testing
- When to retest, the appropriate amount of retesting, and how to evaluate retest results
- Limited circumstances in which resampling is acceptable
- When remixing is allowed
- The significance of averaging test results, and when this practice is acceptable
- Standards for batch release
- The importance of blend testing
- Appropriate sample size, number of samples, and sample location

The *Barr* decision generated considerable debate within the pharmaceutical industry. Soon after the court issued its opinion, FDA announced that FDA investigators would carry the *Barr* decision with them when inspecting companies.[242] Thus, for many years, CGMP standards created by a judge, not scientists, with input from only one company and two FDA district offices, essentially governed proper laboratory and manufacturing practices for the entire industry. FDA adopted many of the court's holdings by issuing guidance that was virtually identical to the opinion.[243]

A few of the court's findings were problematic. Seemingly plausible on paper, they created unnecessary confusion in the laboratory. For example, although the court accepted the use of outlier tests to discard microbiological assay results, the court refused to allow outlier testing for chemical assays. The court's rationale was that because USP did not discuss outliers in chemical testing, USP did not contemplate their use.[244] The court stated that "[t]he substantial variability of microbiological assays supports this distinction. Chemical assays are considerably more precise than biological and microbiological assays, since only the latter testing is 'subject to whims of microorganisms.'"[245]

---

[240] *Id.*

[241] *Id.* at 464.

[242] FDA Inspectors Will Carry Judge Wolin's Decision, Dickinson's FDA, Mar. 15, 1993.

[243] In July 1993, FDA's Mid-Atlantic region formally adopted the court's decision by issuing its "Guide to Inspection of Quality Control Laboratories." With few exceptions, the Guide mirrored the opinion.

[244] Barr Labs., 812 F. Supp. at 469.

[245] *Id.*

First, as Barr demonstrated at trial, the application of outlier tests to chemical assay data was acceptable and standard throughout the industry.[246] Second, as Barr's expert statistician explained in a subsequent commentary on the *Barr* case, the validity of outlier testing does not depend on variability, because variability is taken into account in the test. Thus, an assay with large variability, such as a microbiological assay, would have to show considerable divergence due to the suspected outlier for the value to be rejected. Because of lower variability, testing for an outlier in a chemical assay might reject a less distant observation.[247]

Third, nothing in the USP suggested that silence on a point was prohibitory. In fact, the USP implied that the use of scientifically sound statistical principles was acceptable:

> Confusion of compendia standards with release tests and statistical sampling plans occasionally occurs. Compendial standards define what an acceptable article is and give test procedures that demonstrate that the article is in compliance. *** The manufacturer's release specifications, and compliance with good manufacturing practices generally, are developed and followed to assure that the article will indeed comply with compendial standards until its expiration date.
>
> ***
>
> [USP tests on multiple dosage units] should not be confused with statistical sampling plans. *** Repeats, replicates, or extrapolations of results to larger populations are neither specified nor proscribed by the [USP] compendia; such decisions are dependent on the objectives of the testing. Commercial or regulatory compliance testing, or manufacturer's release testing, may or may not require examination of additional specimens, in accordance with predetermined guidelines or sampling strategies.[248]

Even FDA's statistical expert[249] admitted that a firm could distinguish between an OOS result caused by the "random variability inherent in the data" and an OOS result that is the "result of gross deviation from prescribed experimental procedure."[250] Faced with a gross deviation, FDA's expert stated, a statistical outlier test could be a useful tool in a laboratory investigation.[251]

Soon after the court issued the *Barr* decision, the USP Committee of Revision proposed a change to the General Notices section of the USP to indicate that the USP neither supported nor prohibited the use of outlier testing.[252] The USP adopted the proposed change to the General Notices section in USP 23, which became effective January 1, 1995.[253] Thus, almost two years elapsed before the outlier testing issue was clarified by USP.

---

[246] Transcript, *supra* note 176, at 1394-95.

[247] Sanford Bolton, Should a Single Unexplained Failing Assay be Reason to Reject a Batch?, 10 CLINICAL RES. AND REG. AFF. 159, 174 (1993).

[248] United States Pharmacopeia 9 (23d ed. 1995).

[249] FDA presented no live expert testimony on this issue at trial. In order to rebut Barr's testimony, FDA submitted a post-trial declaration from Dr. Thomas Hammerstrom, a mathematical statistician in the Biometrics Division of FDA's Center for Drug Evaluation and Research.

[250] Hammerstrom Declaration at § 11, United States v. Barr Labs., Inc., 812 F. Supp. 458 (D.N.J. 1993)

[251] *Id.*

[252] See 19 Pharmacopieal Forum 5179, 5193 (1993) ("Repeats, replicates, statistical rejection of outliers, or extrapolations of results to larger populations are neither specified nor proscribed by the compendia; such decisions are dependent on the objectives of the testing." The General Notices section applies to all standards, tests, and assays set forth in the USP. Moreover, if a specific section of USP is silent on a particular issue, the General Notices section applies).

[253] United States Pharmacopeia 6-10 (23d ed. 1995).

A second source of confusion was the court's ruling on retests. As discussed above, the court stated that firms must do "enough" testing to reach a conclusion to confirm or reject an OOS result. The court then surmised that if a laboratory obtained seven out of eight passing results that might be sufficient. Neither party proffered evidence indicating that seven-out-of-eight testing would be an acceptable practice. In effect, the court pulled the seven-of-eight formula out of a hat.[254] Firms were left to wonder whether seven out of eight was the governing standard for retests. The seven-of-eight formula raised questions about how much testing is enough, how much is too much and when the amount of testing would cross the line and trigger observations th firm was "testing products into compliance."[255]

A third problem created by the court was its position on averaging. As explained above, the court determined that averaging might mask variability among initial values where some are within specification and some are not. Although the court's point was valid—OOS data cannot be eliminated through averaging—the court's solution was extreme. Just as OOS results give a firm useful information about a batch, an average may give the firm more information about the true value of the batch than will individual results.[256] Moreover, because averaging was prohibited by the Barr opinion, firms were required to devise new release criteria not knowing whether those criteria would be acceptable to FDA.

Another problem created by the court's ruling was that it provided scant information with respect to validation of methods. The decision stated that analytical method validation was required (subject to the exceptions specified in the CGMP regulation), but it said nothing about how to structure validation studies that would comply with CGMP. The court was unable to provide guidance of this nature because the government did not present any evidence at trial to indicate its view of CGMP compliant analytical method validation.

The *Barr* decision provided a framework for CGMP compliance. It also highlighted the need for the agency to provide clear and specific guidance to industry, or risk having U.S. district courts continue to establish new standards for CGMP compliance. In order to address this need, the agency eventually began to develop guidance documents reflecting FDA's "current thinking" on CGMPs. While these guidance documents are

---

[254] In footnote 9 to the opinion, the court indicated that it relied on Compliance Officer Mulligan's testimony to arrive at the seven-out-of-eight standard. Mr. Mulligan's testimony hardly supports the court's conclusion:

Q. If you have one value in and one value out, how would you propose to proceed to determine what the true value is?

A. You would do additional testing on that same bottle.

Q. And how—if you got two more passing results, you now have three passing and one out of specification results, would you then determine the bottle was within specification?

A. It's not—I don't see that as conclusive.

Q. And if you take two more and you have five in spec and one out of spec, would that be conclusive to you, Mr. Mulligan?

A. No, but it's getting there.

Q. And if you took two more and had seven in spec and one out of spec, would that be conclusive to you?

A. I would defer to someone more expert in antibiotics to make that decision.

Transcript, *supra* note 176, at 804-05.

[255] As discussed at the beginning of this chapter, a basic CGMP principle is that quality, safety, and effectiveness must be designed and built into a product. If a test result falls outside predetermined specifications, the manufacturer is required to determine whether the OOS result is valid and, if so, the root cause of the problem. Understanding the root cause of the problem enables the manufacturer to take appropriate corrective and preventive action to ensure that the product is safe and effective. While retesting is appropriate as part of a formal investigation, a manufacturer that retests simply to obtain a favorable result fails to understand the reason for the OOS result or take corrective action to fix the problem. This practice is described as "testing into compliance," and it is a significant departure from CGMPs. As stated in the opinion, "[b]ecause the initial tests are genuine ... additional testing alone cannot infuse the product with quality." United States v. Barr Labs., Inc., 812 F. Supp. 458, 469 (D.N.J. 1993).

[256] See Reply to Plaintiff's Response to Barr Laboratories' Motion for Clarification and/or Recommendation at 6, United States v. Barr Labs., Inc., 812 F. Supp. 458 (D.N.J. 1993) (No. 92-1744).

not legally binding, firms that do not comply with FDA's guidance should follow alternative practices that are scientifically sound and in compliance with CGMPs.

## C. FDA Guidance to Industry

Over the past few years FDA has issued hundreds of guidance documents. These guidances cover topics ranging from scientific issues, evaluation of product approvals and manufacturing practices.[257] They communicate FDA's interpretations of FDCA provisions and related matters, as well as provide compliance frameworks. Typically, FDA will publish a draft guidance document and solicit comments from stakeholders to be considered in the formulation of the final guidance document.[258] As previously mentioned, guidance documents are not legally binding; however, their proliferation underscores their importance to both industry and FDA. This section contains an overview of the FDA guidance documents most pertinent to CGMPs.

### 1. OOS Guidance

In October 2006, FDA issued its final "Guidance for Industry: Investigating Out-of-Specification (OOS) Test Results for Pharmaceutical Production."[259] The Guidance applies to specifications and established acceptance criteria in drug applications, drug master files (DMFs), official compendia, and those established by the product manufacturer.[260] FDA recommends use of the Guidance to evaluate OOS test results for raw materials, components, in-process materials, and finished products.[261]

The Guidance establishes a framework for investigating OOS test results, including the responsibilities of the analyst and laboratory supervisor, requirements for the laboratory phase of the investigation, requirements for the formal investigation extending beyond the laboratory, and the evaluation of test results.[262]

The Guidance document states that an investigation should be conducted whenever an OOS test result is obtained.[263] The purpose of the investigation is to determine the cause of the OOS result and whether the result is associated with other batches of the same drug or other products.[264] FDA notes in the Guidance that rejection of a batch does not eliminate the need for an investigation.[265] Before conducting an investigation, however, the firm must determine whether the OOS result is valid, or is the result of laboratory error.[266]

In order to determine whether there has been a laboratory error, the Guidance recommends an initial assessment of the accuracy of the lab's data, which should be conducted, if possible, before test preparations are

---

[257] Food & Drug Admin., Fact Sheet: FDA Good Guidance Practices (2011), http://www.fda.gov/downloads/AboutFDA/Transparency/TransparencyInitiative/UCM285344.pdf.

[258] *Id.*

[259] Food & Drug Admin., Guidance for Industry: Investigaing Out-of-Specification (OOS) Test Results for Pharmaceutical Production (2006), http://www.fda.gov/downloads/Drugs/Guidances/ucm070287.pdf [Investigating OOS].

[260] *Id.* at 1.

[261] *Id.* at 1-2.

[262] *Id.* at 3-14.

[263] *Id.* at 3.

[264] *Id.*

[265] *Id.*

[266] *Id.*

discarded.[267] If clear evidence of laboratory error exists, the test results should be invalidated. If it is not clear that a lab error occurred, a full-scale OOS investigation should begin.[268]

With regard to retesting, the Guidance states that "[t]he sample used for the retesting should be taken from the same homogeneous material that was originally collected from the lot, tested and yielded the OOS results. For a liquid, it may be from the original unit liquid product or composite of the liquid product; for a solid it may be an additional weighing from the same sample composite prepared for the original test."[269]

With regard to resampling, the Guidance states that resampling "involves analyzing a specimen from any additional units collected as part of the original sampling procedure or from a new sample collected from the batch, should that be necessary."[270] The Guidance states further that the original sample should be sufficiently large to accommodate additional testing, although it may be appropriate to collect a new sample from the batch.[271] Control mechanisms for examination of additional specimens should be in accordance with predetermined procedures and sampling strategies.[272] The Guidance states further that "if the investigation determines that the initial sampling method was inherently inadequate, a new accurate sampling method must be developed, documented, and reviewed and approved by the QCU." [273]

The Guidance makes clear that testing into compliance is not allowed. [274] In what must have been a tremendous relief to industry, FDA implicitly rejected the *Barr* "7 out of 8" testing standard, stating that "the maximum number of retests to be performed on a sample should be specified in advance in a written standard operating procedures (SOP)."[275] FDA stated further that the number of retests could vary, depending on the variability of the test method, as long as the number of retests was based on scientifically sound principles.[276] The Guidance states that "[t]he number of retests should not be adjusted depending on the results obtained, [and] [t]he firm's predetermined testing procedures should contain a point at which the additional testing ends and the batch is evaluated."[277]

With regard to averaging, the Guidance states that when a sample is assumed to be homogeneous, using averages can provide a more accurate result.[278] "[A]n HPLC assay result may be determined by averaging the peak responses from a number of consecutive, replicate injections from the same preparation . . . . The assay result would be calculated using the peak response average. This determination is considered one test and one result."[279] The Guidance does not allow averaging where the test is intended to measure variability in the product, such as blend uniformity or content uniformity testing.[280] Moreover, the Guidance states that averag-

---

[267] *Id.*
[268] *Id.*
[269] *Id.* at 7-8.
[270] *Id.* at 9.
[271] *Id.*
[272] *Id.*
[273] *Id.*
[274] *Id.* at 8.
[275] *Id.*
[276] *Id.*
[277] *Id.*
[278] *Id.* at 9.
[279] *Id.* at 10.
[280] *Id.* at 10-11.

ing the results of an initial OOS result and the additional retest or resample results obtained during the OOS investigation is not appropriate.[281]

In another departure from the *Barr* decision's flat prohibition regarding outlier testing, the Guidance explains that outlier tests may be used in certain circumstances.[282] For "validated chemical tests with relatively small variance, and if the sample being tested can be considered homogeneous . . . an outlier test is only a statistical analysis of the data obtained from testing and retesting. It will not identify the cause of an extreme observation, and therefore, should not be used to invalidate the suspect result . . . [A]n outlier test may be of some value in estimating the probability that the OOS result is discordant from a data set, and this information can be used in an auxiliary fashion, along with all the other data from the investigation, to evaluate the significance of the result."[283]

Finally, while the Barr decision stated that all investigations were required to be completed within 30 days, the language in the Guidance is much more flexible, stating simply that "the investigation should be thorough, timely, unbiased, well-documented, and scientifically sound."[284]

## 2. Economically Motivated Adulteration

### DEG in Glycerin

In May 2007, FDA issued "Guidance for Industry: Testing of Glycerin for Diethylene Glycol."[285] FDA issued this Guidance to alert manufacturers, compounders, repackagers and suppliers about contamination of glycerin with diethylene glycol (DEG), a poison that can be fatal if ingested.[286] Reports of DEG poisoning indicated that consumers of liquid drug products, such as cough syrup and acetaminophen syrup, were affected.[287] FDA determined that there were certain similarities in each case of DEG poisoning.[288] Manufacturers had not performed full identity testing on the glycerin and had relied instead on the supplier's certificate of analysis (COA).[289] Moreover, the chain of custody, or distribution history, could not be established because the glycerin had been sold several times since its manufacture and before its sale to the liquid drug product manufacturer.[290]

In the Guidance, FDA urges that when manufacturers conduct identity testing on incoming glycerin, as required by 21 C.F.R. § 211.84(d)(1), they include the limit test for DEG contained in the USP monograph for glycerin.[291] The Guidance indicates that manufacturers may use an equivalent method, provided that it meets the safety limit for DEG (0.1 percent).[292] The Guidance also includes a recommendation that

---

[281] *Id.*

[282] *Id.* at 11.

[283] *Id.* at 12.

[284] *Id.* at 3.

[285] Food & Drug Admin., Guidance for Industry: Testing of Glycerine Diethylene Glycol (2007), http://www.fda.gov/downloads/Drugs/GuidanceComplianceRegulatoryInformation/Guidances/UCM070347.pdf [Testing Glycerine].

[286] *Id.* at 1.

[287] *Id.*

[288] *Id.* at 2.

[289] *Id.*

[290] *Id.*

[291] *Id.*, at 2-3. As described in the Guidance, the USP monograph contains a two-part identity test. The first test uses infrared absorption and the second test uses gas chromatography.

[292] *Id.*

manufacturers "know their supply chain for glycerin."[293] As discussed earlier, section 711 of FDASIA makes this mandatory.

**Melamine in At-Risk Materials**

In August 2009, FDA issued "Guidance for Industry: Pharmaceutical Components at Risk for Melamine Contamination."[294] FDA issued this Guidance after receiving reports from China of thousands of injuries and several deaths associated with melamine contamination of food.[295] Although it had received no reports of drug products contaminated with melamine, there were many reports of pet illness and deaths due to melamine contaminated pet food in 2007.[296] For these reasons, the agency issued the Guidance as a preventive measure.[297]

Citing 21 C.F.R. § 211.84, which requires all components, including raw materials, to be tested before being released for use in manufacturing, FDA stated that it is critical for manufacturers to identify components that are "at risk" of melamine contamination.[298] The Guidance provides a list of examples of at-risk components, and explains how such components are identified.[299] The Guidance provides a link to assay methods for measuring melamine in foods.[300] Finally, the Guidance recommends that manufacturers "know and monitor their supply chain for any at risk components."[301] As stated above, section 711 of FDASIA makes supply chain management mandatory.[302]

**OSCS in Crude Heparin**

In February 2012, FDA issued "Guidance for Industry: Heparin for Drug and Medical Device Use: Monitoring Crude Heparin for Quality."[303] In 2008, FDA began receiving reports of serious adverse reactions, some of which were fatal, in dialysis patients.[304] After an extensive investigation, FDA determined that batches of heparin API manufactured in China were contaminated with oversulfated chondroitin sulfate (OSCS).[305] Moreover, the problem was not limited to crude heparin in the United States.[306] At least 10 other countries were affected.[307] The agency concluded that the contamination of crude heparin was intentional, in order to reduce costs of production.[308] Another form of economically motivated adulteration described in the Guidance is use of bovine material instead of porcine intestinal mucosa, the only material approved in the United States for use

---

[293] *Id.* at 3.

[294] FOOD & DRUG ADMIN., GUIDANCE FOR INDUSTRY: PHARMACEUTICAL COMPONENTS AT RISK FOR MELAMINE CONTAMINATION (2009), http://www.fda.gov/downloads/Drugs/GuidanceComplianceRegulatoryInformation/Guidances/UCM175984.pdf [MELAMINE CONTAMINATION].

[295] *Id.* at 1.

[296] Food & Drug Admin., Melamine Pet Food Recall of 2007, http://www.fda.gov/AnimalVeterinary/SafetyHealth/RecallsWithdrawals/ucm129575.htm (last visited April 4, 2013).

[297] MELAMINE CONTAMINATION, *supra* note 285, at 2.

[298] *Id.* at 3 (citing 21 C.F.R. § 211.84).

[299] *Id.* at 4-5.

[300] *Id.* at n.6.

[301] *Id.* at 3.

[302] See *supra* text accompanying notes 66-74.

[303] FOOD & DRUG ADMIN., GUIDANCE FOR INDUSTRY: HEPARIN FOR DRUG AND MEDICAL DEVICE USE: MONITORING CRUDE HEPARIN FOR QUALITY (2012), http://www.fda.gov/downloads/Drugs/GuidanceComplianceRegulatoryInformation/Guidances/ UCM291390.pdf [CRUDE HEPARIN].

[304] *Id.* at 2.

[305] *Id.*

[306] *Id.*

[307] *Id.*

[308] *Id.*

in crude heparin.[309] The substitution of bovine material creates the risk of contamination with the agent that causes bovine spongiform encephalopathy (BSE).[310]

In the Guidance, FDA recommends that drug and medical device manufacturers test each shipment of crude heparin to confirm the species of origin and to determine whether the shipment is contaminated with OSCS.[311] The Guidance contains a reference to an assay method that measures OSCS using Strong Anion Exchange high-pressure liquid chromatography.[312] An alternative method can be used, provided that it has been qualified for use in screening crude heparin for OSCS.[313] The Guidance urges manufacturers to "know their suppliers."[314] As discussed previously, this requirement is now mandatory for drug manufacturers.[315] Manufacturers are encouraged to employ the controls described in ICH Q7 (described below) and to conduct thorough and timely investigations of deviations and quality failures.[316] Additionally, manufacturers are to reject contaminated heparin and notify the local FDA district of the contamination.[317]

### 3. Quality Systems Approach to CGMP Regulation

In September 2006, FDA issued "Guidance for Industry: Quality Systems Approach to Pharmaceutical CGMP Regulations."[318] The agency developed this Guidance "as a bridge between the 1978 regulations and our current understanding of quality systems," with the following goals: reducing the number of nonconforming products entering interstate commerce, harmonizing CGMP regulations with other quality management systems such as ISO 9000, non-U.S. systems and FDA's quality system regulations for medical devices, lowering the risk of manufacturing problems, which could result in less frequent and shorter inspections; and providing a framework for building quality into products throughout the life cycle (quality by design).[319]

The Guidance describes the following key quality concepts critical to an understanding of quality systems:

- Products and manufacturing processes must be designed to ensure that products consistently meet predetermined specifications.[320]
- Quality risk management enables manufacturers to more easily establish specifications and process parameters for manufacturing, evaluate risks associated with changes, conduct investigations, and establish corrective actions.[321]

---

[309] Crude Heparin, *supra* note 303, at 3.

[310] *Id.*

[311] *Id.* at 5.

[312] *Id.*

[313] *Id.*

[314] *Id.* at 5.

[315] Supply chain management requirements for medical device manufacturers are well-established under FDA's quality system regulation purchasing controls. See 21 C.F.R. § 820.50.

[316] Crude Heparin, *supra* note 303, at 5.

[317] *Id.* at 6.

[318] Food & Drug Admin., Guidance for Industry: Quality Systems Approach to Pharmaceutical Current Good Manufacturing Practice Regulations (2006), http://www.fda.gov/downloads/Drugs/GuidanceComplianceRegulatoryInformation/ Guidances/ UCM070337.pdf [Quality Systems].

[319] *Id.* at 3.

[320] *Id.*

[321] *Id.* at 5.

- Manufacturers must conduct thorough root cause analysis to determine why a problem occurred, and develop corrective and preventive actions designed to address the root cause and ensure that the problem does not recur.[322]
- Through change control, manufacturers are able to evaluate proposed changes and ensure that the changes do not have unintended consequences.[323]

The Guidance describes the responsibilities of the quality unit and emphasizes that this unit should be independent from other business units.[324] In limited circumstances in which a single individual performs quality and production functions, the Guidance makes clear that the person is still accountable for "implementing all the controls and reviewing the results of manufacture to ensure that product quality standards have been met."[325] In such cases, another qualified individual who is not involved in manufacturing should review quality unit activities on a periodic basis.[326]

The quality systems model outlined in the Guidance contains four major components: 1) management responsibility; 2) resources; 3) manufacturing operations; and 4) evaluation activities.

**Management Responsibility**

The Guidance provides recommendations to the pharmaceutical industry that are consistent with the requirements in the quality system regulations (QSRs) for medical devices.[327] The Guidance states that management bears responsibility for the success of the quality system. Senior management "should demonstrate commitment to developing and maintaining their quality system."[328] In addition, management should ensure that the quality system is in alignment with the manufacturer's mission and strategic plans.[329] Senior management should take an active role in the design, implementation and monitoring of the quality system, and should provide visible support to the quality system, ensuring, for example, that the quality unit has equal standing with other business units and encouraging communication on quality issues at all levels within the firm.[330] Management is responsible for ensuring that "assigned authorities and responsibilities support the production, quality, and management activities needed to produce quality products," and for clearly articulating roles, responsibilities, and authorities.[331] The individual appointed to manage the quality system must be given the authority to identify problems and to take corrective actions.[332] The quality system should be designed to provide "clear organizational guidance" and enable issues to be evaluated in a systematic fashion.[333] The quality system should also take into consideration all outsourced operations.[334]

---

322 *Id.*

323 *Id.*

324 Quality Systems, *supra* note 318, at 5-6.

325 *Id.* at 6.

326 *Id.*

327 See *id.* at 3. The quality system regulations for medical devices can be found at 21 C.F.R. pt. 820.

328 Quality Systems, *supra* note 318, at 8.

329 *Id.*

330 *Id.* at 9.

331 *Id.*

332 *Id.*

333 Quality Systems, *supra* note 318, at 9.

334 *Id.*

Senior managers should review the performance of the quality system on a regular schedule.[335] These "management reviews" should include an evaluation of quality metrics such as internal audit findings, complaints, trends, and CAPAs.[336] Management reviews should be documented, and planned actions should be implemented through CAPAs or change controls.[337]

**Resources**

The second key component of a robust quality system described in the Guidance is the allocation of sufficient resources to manufacture products that consistently meet predetermined specifications and quality standards.[338] The Guidance emphasizes that the development of highly qualified and trained personnel is essential.[339] The Guidance explains that in addition to being trained on specific tasks and regulatory requirements, personnel should understand the significance of their activities with respect to the product, and the patient or consumer.[340] In addition, FDA recommends that training programs be designed to identify training needs, provide appropriate training, evaluate the effectiveness of training[341] and require complete and accurate documentation of training and retraining.[342]

The Guidance states that outsourced operations are an extension of the manufacturer's own operations, and that ensuring the contractor's compliance with CGMPs is the manufacturer's responsibility.[343] The Guidance describes the need for qualification and monitoring of suppliers.[344] Citing 21 C.F.R. § 211.22(a), the Guidance reminds the manufacturer that its quality unit is responsible for approving or rejecting materials and services provided under contract.[345]

**Manufacturing**

With respect to manufacturing, the Guidance states that "significant characteristics of the product being manufactured should be defined from design to delivery, and control should be exercised over all changes."[346] The Guidance also discusses the need for documentation of processes, controls, and changes.[347]

The Guidance states that a "robust quality system will ensure that all inputs to the manufacturing process are reliable because quality controls will have been established for the receipt, production, storage, and use of all inputs," and then provides specific guidance on the use of COAs and testing of incoming materials.[348] Adopting a concept from the medical device QSRs, the Guidance states that "[a]s an essential element of purchasing controls,

---

335 *Id.* at 10. In general, manufacturers conduct these reviews on an annual or even quarterly basis. The frequency of management reviews will depend on a manufacturer's state of control. For example, significant audit findings, increased recalls or an FDA Warning Letter signal a need for more frequent management reviews.

336 *Id.* at 10-11.

337 *Id.* at 11.

338 Quality Systems, *supra* note 318, at 12.

339 *Id.* at 13.

340 *Id.*

341 The guidance does not explain how training program effectiveness should be evaluated, but in practice, this should include evaluation of program structure, specific training materials and the effectiveness of trainers. Training effectiveness should be evaluated as appropriate during root cause analysis.

342 Quality Systems, *supra* note 318, at 13.

343 *Id.* at 14.

344 *Id.*

345 *Id.* (citing 21 C.F.R. § 211.22(a)).

346 *Id.* at 15.

347 *Id.*

348 Quality Systems, *supra* note 318, at 16.

it is recommended that data trends for acceptance and rejection of materials be analyzed for information on supplier performance."[349] The Guidance provides recommendations for periodic risk-based audits of suppliers, and also recommends that supplier changes be governed by the manufacturer's change control system.[350]

The Guidance also provides recommendations related to design control and risk management.[351] With respect to process validation, the Guidance states that "[a]lthough initial commercial batches can provide evidence to support the validity and consistency of the process, *the entire product life cycle should be addressed* by the establishment of continual improvement mechanisms in the quality system. Thus . . . process validation is not a one-time event, but an activity that continues throughout a product's life."[352]

The Guidance provides recommendations for monitoring of critical processes, and reminds industry of the requirements for second-person verification (or a validated computer system), concurrent documentation, and control of in-process materials.[353] The Guidance recommends that manufacturers develop procedures for monitoring, measuring, and analyzing operations, through the use of analytical methods and/or statistical analysis.[354] In addition, the Guidance states that trends "should be continually identified and evaluated."[355] The Guidance explains that trend data can identify potential variances (sometimes referred to as "process drift") before a problem occurs.[356]

The Guidance states that "a key component in any quality system is handling nonconformities and/or deviations."[357] The Guidance states that discrepancies should be documented appropriately and should be investigated if they have the potential to affect product quality.[358] The Guidance further states that a firm's procedures should establish roles and responsibilities for operational decisions related to the discovery of a nonconformity or deviations (e.g., halting production), documenting the problem, conducting an investigation, and taking corrective action.[359]

**Evaluation Activities**

1. Monitoring Trends and Improving Systems

With respect to evaluation activities, the Guidance reiterates the need for continuous monitoring of trends.[360] Through the collection and analysis over time of data generated during manufacturing, from complaints and other activities, manufacturers are able to determine whether processes are in a state of control or are drifting out of control.[361] The Guidance describes trending as an early warning system for detection of

---

[349] *Id.* (emphasis added).

[350] *Id.* at 16-17.

[351] *Id.* at 17.

[352] *Id.* (emphasis in original). In the Guidance it is stated that even with excellent designs "initial *conformance batches* will provide confidence that future batches will meet specifications only if the process is repeated within defined operating parameters, equipment tolerances, personnel practices, environmental attributes, and material quality." *Id.* at n.15 (emphasis in original).

[353] *Id.* at 18.

[354] Quality Systems, *supra* note 318, at 18.

[355] *Id.* at 19.

[356] *Id.*

[357] *Id.*

[358] *Id.* at 19-20. This is also required under 21 C.F.R. § 211.192.

[359] *Id.* at 20-21.

[360] Quality Systems, *supra* note 318, at 21.

[361] *Id.*

problems, and development of corrective and preventive actions.[362] The Guidance states that trending can also be used to plan internal audits.[363]

2. Internal Audits

The Guidance also recommends that firms conduct internal audits on a periodic, scheduled basis.[364] The Guidance states that procedures should describe the required training for auditors, their responsibilities, and how audits are to be conducted. Procedures should also define the audit's scope, audit plans, and the audit process.[365] The Guidance states further that "[i]t is critical to maintain records of audit findings and assign responsibility for follow-up to prevent problems from recurring."[366] To provide further encouragement to firms to conduct frequent audits and document them appropriately, the Guidance reminds industry of its policy not to review or copy any documentation resulting from internal audits.[367]

3. Risk Management

The Guidance reiterates the need for quality risk management, which includes "assessing the risks, selecting and implementing risk management controls commensurate with the level of risk, and evaluating the results of the risk management efforts."[368] The Guidance states that proper quality risk management helps control and manage change.[369]

4. Corrective and Preventive Actions

The Guidance provides FDA's thinking and recommendations related to corrective actions.[370] The Guidance describes corrective action as a "reactive tool," presumably in contrast to more active tools such as design control, purchasing control, process control, monitoring, and quality risk management.[371] The Guidance states that corrective action procedures should ensure that "the need for action is evaluated relevant to the possible consequences, the root cause of the problem is investigated, possible actions are determined, a selected action is taken within a defined timeframe, and the effectiveness of the action taken is evaluated. It is essential to document corrective action taken."[372] The Guidance also explains that it is "essential" for corrective actions to reduce the possibility that the problem will recur, and provides examples of sources of quality data that should be evaluated, such as nonconformance reports, complaints, and audits, as part of the CAPA process.[373] The Guidance also explains the need to evaluate the effectiveness of preventive actions.[374] The Guidance concludes by stating that a successful quality system is characterized by science-based decisionmaking, an understanding of all aspects of the product (intended use, product lifecycle, potential for process weakness), well-defined systems related to control and investigation of deviations, monitoring, and evaluation of product quality and management

---

[362] *Id.*

[363] *Id.*

[364] *Id.* at 21-22.

[365] *Id.* at 22.

[366] *Id.*

[367] *Id.* This policy was announced in FDA's CPG Sec. 130.300. FOOD & DRUG ADMIN., Sec. 130.300 FDA Access to Results of Quality Assurance Program Audits and Inspections, in COMPLIANCE POLICY GUIDE (2007), http://www.fda.gov/ICECI/ComplianceManuals/CompliancePolicyGuidanceManual/ucm073841.htm [Sec. 130.300]. The policy does not apply in directed or for-cause inspections. *Id.*

[368] Quality Systems, *supra* note 318, at 22.

[369] *Id.*

[370] *Id.* at 22-23.

[371] *Id.* at 22.

[372] *Id.*

[373] *Id.* at 22-23.

[374] Quality Systems, *supra* note 318, at 23.

support.[375] The Guidance suggests that the adoption of a robust quality system is not only good manufacturing practice, but also good business practice.[376]

## 4. Process Validation

In January 2011, FDA issued "Guidance for Industry: Process Validation: General Principles and Practices" (the 2011 Guidance).[377] The 2011 Guidance revised and replaced FDA's earlier guidance on process validation, "Guideline on General Principles of Process Validation," issued in May 1987.[378] The 2011 Guidance reflects goals established in FDA's "Pharmaceutical CGMPs for the 21st Century."[379]

The 2011 Guidance defines process validation as "the collection and evaluation of data, from the process design stage through commercial production, which establishes scientific evidence that a process is capable of consistently delivering quality products. Process validation involves a series of activities taking place over the lifecycle of the product and process."[380] The 2011 Guidance describes three stages of process validation: Stage 1, Process Design; Stage 2, Process Qualification; and Stage 3, Continued Process Verification. In this document, FDA recommends an "integrated team approach" to process validation, with evaluation taking place throughout the product lifecycle. FDA states further that attributes and parameters should be evaluated based on their impact on the product, and should be reevaluated as new information becomes available. The agency states that "[h]omogeneity within a batch and consistency between batches are goals of process validation activities."[381]

In Stage 1 – Process Design, product development includes establishment of the product dosage form, quality attributes, and general manufacturing process. In this stage, firms should identify factors that could contribute to variability in the process, such as component lot variability, different production operators, or changes in environmental conditions. Also during this stage, firms should develop strategies to reduce input variation.[382]

In Stage 2 – Process Qualification, the first element consists of qualification of utilities and equipment to ensure that they are appropriate for their intended use and perform as expected. These qualification activities should be described in a written project plan, and the results should be summarized and evaluated in a written report. The firm's quality control unit must review and approve the plan and report.[383]

The second element in Stage 2 is process performance qualification (PPQ). The PPQ involves studies that combine the qualified utilities and equipment with trained personnel, the commercial manufacturing process, control procedures and components to produce commercial-scale batches. The goal of the PPQ is to establish

---

[375] *Id.* at 24.

[376] *Id.*

[377] Food & Drug Admin., Guidance for Industry: Process Validation: General Principles and practices (2011), http://www.fda.gov/downloads/Drugs/GuidanceComplianceRegulatoryInformation/Guidances/ucm070336.pdf [2011 Guidance]. The 2011 Guidance covers drugs and biologics, and applies to finished products, APIs, and the drug constituent of a combination product.

[378] Food & Drug Admin., Guideline on General Principles of Process Validation (1987).

[379] 2011 Guidance, *supra* note 377, at 3.

[380] *Id.* at 4.

[381] *Id.* at 7.

[382] *Id.* at 8-9.

[383] *Id.* at 10-11.

that the manufacturing process for commercial batches performs as expected. In general, the PPQ will require more extensive sampling, testing, and evaluation than that required for commercial production.[384]

The PPQ must be conducted pursuant to a written protocol that specifies manufacturing conditions, such as operating parameters, process limits and inputs, controls, testing to be performed, such as in-process and release testing, and anticipated outcomes. The protocol must be reviewed and approved by operational units and quality. Note that the PPQ should not be executed until the protocol is approved. Protocol deviations are not allowed except in accordance with established SOPs, or as provided in the protocol.[385] The PPQ lots must be manufactured according to the manufacturing process and procedures to be used during commercial manufacturing.[386]

The firm should prepare a PPQ report after execution of the protocol is complete. The report should be comprehensive, with an analysis of all protocol steps, data collected, unexpected observations, deviations, corrective actions required, and a conclusion whether the process met all criteria established in the protocol and is in a state of control. The report must be reviewed and approved by operational units and quality.[387]

In Stage 3 – Continued Process Verification, the firm must ensure that the process remains in a state of control throughout the product lifecycle. This requires ongoing data collection and evaluation of trends related to product quality. The 2011 Guidance recommends the use of statistical process control techniques to ensure that the commercial manufacturing process remains in a state of control.[388] FDA also recommends that procedures "describe how trending and calculations are to be performed and should guard against overreaction to individual events and against the failure to detect unintended process variability."[389]

The agency notes that firms should evaluate product complaints, OOS results, deviation reports, adverse event reports, and other quality data sources to evaluate process variation. In addition, changes should be controlled to ensure that there are no unintended effects on product quality.[390]

The 2011 Guidance emphasizes that in most cases, the PPQ must be complete before distribution of product. In "special situations," the PPQ protocol may be designed to allow concurrent release of batches, i.e. before complete execution of the protocol.[391] The agency indicates that concurrent release "might" be appropriate for product manufactured infrequently, such as orphan drugs, products with a short half-life, such as radiopharmaceuticals, and medically necessary drugs that are being manufactured in coordination with FDA to address a drug shortage.[392]

Technical information related to the 2011 Guidance is included in Appendix A and Appendix B of the Primer.

---

[384] *Id.* at 11-12.

[385] 2011 Guidance, *supra* note 369, at 13.

[386] *Id.* at 12-13.

[387] *Id.* at 13-14.

[388] Statistical process control techniques can be useful in identifying inter-batch or intra-batch variability. They can also be used to detect process drift. For example, if the acceptable finished product composite assay range is 90.0 to 110.0 percent, a downward trend in assay values could indicate a problem with the manufacturing process.

[389] 2011 Guidance, *supra* note 377, at 14-15.

[390] *Id.* at 15.

[391] *Id.* at 16.

[392] *Id.*

## 5. ICH Guidance Documents

The International Conference on Harmonisation of Technical Requirements for Registration of Pharmaceuticals for Human Use (ICH) represents an effort among various national regulators to develop uniform standards for safety, quality and efficacy of pharmaceutical products.[393] The ICH developed in response to the globalization of the drug industry, which coincided with a rapid growth in regulatory regimes.[394] During the 1980s, the European Community moved toward establishing a single approval pathway for pharmaceutical products.[395] The success of the one European approval pathway would foster cross-collaboration among regulators in Europe, the United States and Japan, seeking to achieve greater international harmonization. Eventually, the success of their efforts would result in various ICH Guidelines.[396] FDA has issued its own ICH Guidance documents largely incorporating the principles of the ICH Guidelines. Below is a summary of key FDA ICH Guidance Documents related to quality in pharmaceuticals.

### ICH Q7

In order to ensure that quality is built into a finished drug product even before the manufacturing process for that finished drug product begins, FDA has issued guidance on CGMPs for Active Pharmaceutical Ingredients (APIs).[397] FDA has long recognized that CGMP requirements are applicable to APIs, noting in the Compliance Program Guidance Manual that "[n]o distinction is made between an API and a finished pharmaceutical in the [FDCA]."[398] The Q7 Guidance follows a standard format for CGMPs, and includes production and in-process controls, validation, laboratory controls, and quality management.[399] In addition, the Q7 Guidance highlights recordkeeping, including traceability of distributed APIs.[400] Lastly, the Q7 Guidance contains recommendations especially applicable to API usage in clinical trials.[401]

### ICH Q8

The ICH Q8 Guidance discusses the recommended contents to be included under Module 3, specifically pharmaceutical development, of the Common Technical Document (CTD).[402] The CTD is a format developed to harmonize some aspects of drug product submissions, or modules, so they can be uniformly reviewed by various national regulatory bodies.[403] Pharmaceutical development concerns itself with the design of product quality.[404] As part of the CTD submission of pharmaceutical development, the Guidance advises identification and control of processes that are critical to product quality.[405] The Q8 Guidance provides recommendations

---

[393] ICH, About ICH – History, http://www.ich.org/about/history.html (last visited April 4, 2013).

[394] *Id.*

[395] *Id.*

[396] *Id.*

[397] FOOD & DRUG ADMIN., GUIDANCE FOR INDUSTRY: Q7A GOOD MANUFACTURING PRACTICE GUIDANCE FOR ACTIVE PHARMACEUTICAL INGREDIENTS (2001), http://www.fda.gov/iceci/compliancemanuals/compliancepolicyguidancemanual/ucm200364.htm [Q7 GUIDANCE].

[398] See FOOD & DRUG ADMIN., Program No. 7356.002F Active Pharmaceutical Ingredient (API) Process Inspection, in COMPLIANCE PROGRAM GUIDANCE MANUAL (2010), http://www.fda.gov/downloads/ICECI/ComplianceManuals/Compliance ProgramManual/UCM125420.pdf. See FOOD & DRUG ADMIN., Program No. 7346.832 Pre-Approval Inspections/ Investigations 4, in COMPLIANCE PROGRAM GUIDANCE MANUAL (2010), http://www.fda.gov/downloads/Drugs/ DevelopmentApprovalProcess/Manufacturing/ QuestionsandAnswersonCurrentGoodManufacturingPracticesCGMPforDrugs/ UCM071871.pdf.

[399] Q7 GUIDANCE, *supra* note 397, at ii.

[400] *Id.* at 15-19, 39-41.

[401] *Id.* at 44-46.

[402] FOOD & DRUG ADMIN., GUIDANCE FOR INDUSTRY: Q8 (R2) PHARMACEUTICAL DEVELOPMENT 1 (2006), http://www. fda.gov/downloads/ Drugs/GuidanceComplianceRegulatoryInformation/ Guidances/ucm073507.pdf [Q8 GUIDANCE].

[403] ICH, The Common Technical Document, http://www.ich.org/products/ctd.html (last visited April 4, 2013).

[404] Q8 GUIDANCE, *supra* note 402, at 2.

[405] *Id.*

on the CTD's information requirements for components of drug products, drug products, the manufacturing process, container closure systems, microbiological attributes, and compatibility of the drug product with reconstitution diluents.[406]

### ICH Q9

Another FDA effort to change the CGMP compliance landscape can be found in the Quality Risk Management Guidance, Q9 – Quality Risk Management. This Guidance allows for less intense regulatory scrutiny in exchange for quality management procedures that are adopted consistent with the Guidance.[407] The Q9 Guidance was also generated in part to help promote the regulatory harmonization efforts of the ICH.[408] As part of the Guidance, FDA listed a variety of risk management methodologies that could be utilized, such as such as Hazard Analysis and Critical Control Points (HACCP).[409] FDA stated that quality risk management, while not obviating compliance obligations, can serve to provide "greater assurance of a company's ability to deal with potential risks."[410]

### ICH Q10

FDA has also set forth its expectations for senior management and executives in FDA Guidance for Industry: Q10 Pharmaceutical Quality System (Q10).[411] Among other requirements, FDA expects senior management to oversee pharmaceutical quality research, outsourced activities, purchased materials, and the company's quality system. In Q10, FDA states that "[s]enior management has the ultimate responsibility to ensure an effective pharmaceutical quality system is in place to achieve the quality objectives, and that roles, responsibilities, and authorities are defined, communicated, and implemented throughout the company."[412]

Q10 defines the term senior management as "[p]erson(s) who direct and control a company or site at the highest levels with the authority and responsibility to mobilize resources within the company or site."[413] This definition is based on the ISO 9000-2005 definition for "Top Management," which is the "person or group of people who directs and controls an organization . . . at the highest level."[414] As stated in Q10, senior management must establish a quality policy, ensure the quality system is in place and meets quality objectives, manage resources, communicate internally, and conduct management reviews. Specifically, the document provides that management should:

---

[406] *Id.* at 3-11.

[407] Food & Drug Admin., Guidance for Industry: Q9 Quality Risk Management 8 (2006), http://www.fda.gov/ downloads/Drugs/GuidanceComplianceRegulatoryInformation/ Guidances/ucm073511.pdf.

[408] *Id.* at 1 n.1.

[409] *Id.* at 7.

[410] *Id.* at 7-8.

[411] Food & Drug Admin., Guidance for Industry: Q10 Pharmaceutical Quality System (2009), http://www.fda. gov/downloads/Drugs/ GuidanceComplianceRegulatoryInformation/Guidances/ucm073517.pdf [Q10 Guidance].

[412] *Id.* at 5-6 (emphasis in original). The language of the Q10 reflects that found in *United States. v. Park*, which held that responsible corporate officers who fail to establish sufficient controls over drug products can be held criminally responsible. 421 U.S. 658 (1975). As the Court noted, "the [FDCA] imposes not only a positive duty to seek out and remedy violations when they occur but also, and primarily, a duty to implement measures that will insure that violations will not occur. The requirements of foresight and vigilance imposed on responsible corporate agents are beyond question demanding, and perhaps onerous, but they are no more stringent than the public has a right to expect of those who voluntarily assume positions of authority in business enterprises whose services and products affect the health and well-being of the public that supports them." *Id.* at 673. Further discussion of the *Park* doctrine is included in Chapter III.

[413] Q10 Guidance, *supra* note 411, at 5-6.

[414] Int'l Org. for Standardization, ISO 9000-2005, Quality Management Systems – Fundamentals and Vocabulary (2005).

(1) Participate in the design, implementation, monitoring, and maintenance of an effective pharmaceutical quality system;

(2) Demonstrate strong and visible support for the pharmaceutical quality system, and ensure its implementation throughout their organization;

(3) Ensure a timely and effective communication and escalation process exists to raise quality issues to the appropriate levels of management;

(4) Define individual and collective roles, responsibilities, authorities, and interrelationships of all organizational units related to the pharmaceutical quality system. Ensure these interactions are communicated and understood at all levels of the organization. An independent quality unit/structure with authority to fulfill certain pharmaceutical quality system responsibilities is required by regional regulations;

(5) Conduct management reviews of process performance and product quality and of the pharmaceutical quality system;

(6) Advocate continual improvement; and

(7) Commit appropriate resources.[415]

FDA also expects that senior management will oversee all outsourced activities. In Q10, FDA makes clear that these responsibilities:

> extend[] to the control and review of any outsourced activities and quality of purchased materials. The pharmaceutical company is ultimately responsible to ensure processes are in place to assure the control of outsourced activities and quality of purchased materials.[416]
>
> FDA expects pharmaceutical companies to incorporate quality risk management, including:
>
> (a) Assessing prior to outsourcing operations or selecting material suppliers, the suitability and competence of the other party to carry out the activity or provide the material using a defined supply chain (e.g., audits, material evaluations, qualification).
>
> (b) Defining the responsibilities and communication processes for quality-related activities of the involved parties. For outsourced activities, this should be included in a written agreement between the contract giver and contract acceptor.
>
> (c) Monitoring and review of the performance of the contract acceptor or the quality of the material from the provider, and the identification and implementation of any essential improvements.
>
> (d) Monitoring incoming ingredients and materials to ensure they are from approved sources using the agreed supply chain.[417]

---

[415] Q10 GUIDANCE, *supra* note 411, at 6.
[416] *Id.* at 7 (emphasis added).
[417] *Id.*

Importantly, the requirements of (c) and (d) involve continuing obligations to monitor and identify improvements.

### ICH Q11

FDA's latest ICH Quality Guidance relates to "drug substances" (i.e., APIs)[418] and provides insights on quality approach to these drug components. ICH Q11 specifically addresses the manufacturing process development of drug substances as opposed to ICH Q7, which concerns CGMPs for already established APIs.[419] In other words, ICH Q11 can be considered as Guidance for manufacturing associated with clinical development.[420] Proceeding along with ICH harmonization objectives, FDA's Q11 Guidance provides additional clarification on previous ICH Guidances (Q8, Q9, and Q10) as well as information requirements of Module 3 of the CTD.[421] The Q11 Guidance also differentiates between traditional and enhanced approaches to drug product development.[422] The traditional approach is based on "demonstration of process reproducibility and testing to meet established acceptance criteria."[423] The enhanced approach is one where "risk management and scientific knowledge" are used to comprehend processes and operations that impact quality attributes.[424]

## 6. CGMP Requirements During Clinical Development

CGMPs were specifically intended for large commercial production of drug products.[425] That being said, in the 1978 final rule, FDA took the position that CGMP provisions were widely applicable to drug manufacturing including the production of investigational drugs.[426] However, the preamble to the final rule also stated FDA was considering specific regulations to govern drugs in the development process.[427]

FDA subsequently issued a regulation exempting certain Phase I investigational drug products from the CGMPs in 21 C.F.R. part 211.[428] Specifically the regulation and FDA Guidance state that Phase I investigational drugs are not required to comply with 21 C.F.R. part 211 unless the Phase I drug product is already approved for further development (Phase II, III) or approved for marketing.[429] In other words, a Phase I trial using a drug product already approved for Phase II, III or marketing cannot qualify for the CGMP exemption. The exemption also does not apply to PET drugs, human cell, or tissue products regulated solely under the Public Health Service Act.[430] As discussed further below, these modified CGMP standards are necessary as the manufacturing process during clinical development is continuously being refined. This is especially true with respect to validation, as the manufacturing processes are constantly being modified during the development stage.

---

[418] See 21 C.F.R § 314.3(b).

[419] Food & Drug Admin., Guidance for Industry: Q11 Development and Manufacture of Drug Substances (2012), http://www.fda.gov/downloads/Drugs/GuidanceComplianceRegulatoryInformation/Guidances/ucm261078.pdf [Q11 Guidance].

[420] See *infra* section titled "CGMP Requirements During Clinical Development."

[421] Q11 Guidance, *supra* note 419, at 1.

[422] *Id.*

[423] *Id.* at 1-2

[424] *Id.* at 2.

[425] Food & Drug Admin., Guidance for Industry: CGMP for Phase I Investigational Drugs 2 (2008), http://www.fda.gov/downloads/Drugs/GuidanceComplianceRegulatoryInformation/Guidances/ucm070273.pdf [CGMP Guidance].

[426] Current Good Manufacturing Practice in Manufacture, Processing, Packing, and Holding, 43 Fed. Reg. 45,014, 45,029 (Sept. 29, 1978).

[427] *Id.*

[428] 21 C.F.R. § 210.2(c).

[429] *Id.*; CGMP Guidance, *supra* note 425, at 3.

[430] CGMP Guidance, *supra* note 425, at 4.

Prior to the inclusion of the Phase I exemption in 21 C.F.R. part 211, FDA relied upon its Guidance document, "Preparation of Investigational New Drug Products (Human and Animal)," which established CGMP applicability for investigational drug products.[431] In this Guidance FDA provided recommendations on CGMP compliance noting that investigational drugs are not expected to meet the same CGMP objectives as would be expected for commercial manufacture.[432] For example, the guideline noted that during the development stage, product and process controls may be general in nature as they are likely "undergoing considerable refinement."[433] Other topics addressed in this Guidance are control of component, calculation of yield, equipment identification, packaging/labeling, reserve samples, and records retention.[434] This Guidance is still valid for investigational drugs that are not Phase I CGMP exempt drug products.[435]

For Phase I CGMP exempt drug products, FDA has another Guidance document, "CGMPs for Phase I Investigational Drugs," which provides an overview of applicable standards. FDA recommends careful consideration of hazards and risks with Phase I product development especially when manufacture occurs in a laboratory or other non-traditional settings.[436] Some of the recommendations include using streamlined approaches to product development (i.e. utilizing prepackaged and CGMP compliant inputs) and use of disposable equipment (to reduce necessary sanitation).[437] The Guidance addresses a host of CGMP categories such as personnel, QC function, facilities and equipment, control of components, containers, closures, manufacturing records, laboratory controls, packaging, labeling, distribution, and recordkeeping.[438] In addition, the Guidance also addresses special topics and requirements for biological, biotechnological and sterile products.[439]

## 7. Aseptic Processing and Additional CGMP Requirements for Sterile Products and Biologics

In 1987, FDA issued guidance to assist industry in its compliance with CGMPs related to the preparation of sterile drug and biologic products.[440] The Guidance discussed the appropriate environmental controls required for areas at different production stages and also stated the importance of ensuring adequate microbial control over components and containers of sterile products.[441] In addition, the Guidance provided recommendations on laboratory controls and sterility testing.[442] The majority of the document, however, was devoted to production/process controls and validation.[443]

---

[431] FOOD & DRUG ADMIN., CENTER FOR DRUG EVALUATION & RES., GUIDELINE ON THE PREPARATION OF INVESTIGATIONAL NEW DRUG PRODUCTS (HUMAN AND ANIMAL) (1991), http://www.fda.gov/downloads/Drugs/GuidanceComplianceRegulatoryInformation/Guidances/ucm070315.pdf [INVESTIGATIONAL NEW DRUG PRODUCTS]. The exemption regulation was published subsequent to the 1991 guidelines. See Current Good Manufacturing Practice and Investigational New Drugs Intended for Use in Clinical Trials, 73 Fed. Reg. 40,453, 40,462-63 (July 15, 2008).

[432] See generally INVESTIGATIONAL NEW DRUG PRODUCTS, *supra* note 431.

[433] *Id.* at 4.

[434] See generally *id.*

[435] CGMP GUIDANCE, *supra* note 425, at 3.

[436] *Id.* at 6

[437] *Id.*

[438] *Id.* at 7-11.

[439] *Id.* at 12-13.

[440] FOOD & DRUG ADMIN., GUIDANCE FOR INDUSTRY: GUIDELINE ON STERILE PRODUCTS PRODUCED BY ASEPTIC PROCESSING (1987).

[441] *Id.* at 6-19.

[442] *Id.* at 32-41.

[443] *Id.* at 20-32.

In 2002, FDA joined with industry and academia to form a working group to consider aseptic processing.[444] The input of this working group helped FDA formulate a new, updated guidance on aseptic processing in 2004.[445] The 2004 Guidance updated the previous 1987 Guidance with respect to "personnel qualification, cleanroom design, process design, quality control, environmental monitoring, and review of production records."[446] Both the 2004 and previous 1987 Guidance were developed because the 1978 CGMPs on aseptic processing had not been updated to reflect current industry practices.[447]

In addition to the new guidance, FDA included aseptic processing requirements in its 2008 CGMP regulations. FDA stated that the updated CGMPs would serve its harmonization objective and updated the standards to current practices.[448] FDA also stated that the new CGMPs did not affect the 2004 Guidance and that compliance with the Guidance's recommendations would satisfy certain aseptic standards.[449]

Although the 2004 Guidance does not address all aspects of CGMP application to sterile products, it does provide additional information regarding FDA's "current thinking" about aseptic processing. The Guidance clarifies CGMP requirements for sterile drug and biologic products in the following areas: building and facilities, personnel training, component/container/closure controls, endotoxin control, time limitations, process validation, laboratory controls, sterility testing, and batch record review.

**Building and Facilities**

In this category, the Guidance emphasizes the importance of adequate air quality summarizing clean air classifications and appropriate action level for microbiological quality as they relate to clean area control parameters.[450] FDA notes that the two most important areas of concern are the "critical area" and "supporting clean area," each with its own air quality recommended standard.[451] In addition, the Guidance describes appropriate air filtration design, including appropriate membrane and HEPA filter utilization.[452]

**Personnel Training**

The Guidance notes that personnel interaction introduces a risk of compromising sterility and thus a well-designed system for aseptic processing minimizes human operator direct involvement with the sterile product.[453] In addition, the Guidance recommends adequate training, a policy of using sterile instruments to interact with sterile components, appropriate gowning, and a personnel monitoring program.[454]

**Component/Container/Closure Controls**

Aseptic processing sterilizes each component (e.g., API), or combination of components, by determining the microbial content of each input and establishing thresholds for acceptance.[455] The Guidance notes

---

[444] Amendment to Current Good Manufacturing Practice Regulations for Finished Pharmaceuticals, 72 Fed. Reg. 68,064, 68,065 (Dec. 4, 2007).

[445] *Id.*

[446] Food & Drug Admin., Guidance for Industry: Sterile Drug Products Produced by Aseptic Processing – Current Good Manufacturing Practice 5 (2004), http://www.fda.gov/downloads/Drugs/.../Guidances/ucm070342.pdf [Sterile Drug Products].

[447] Amendment to Current Good Manufacturing Practice Regulations for Finished Pharmaceuticals, 72 Fed. Reg. at 68,065.

[448] *Id.*

[449] *Id.*

[450] Sterile Drug Products, *supra* note 446, at 5.

[451] *Id.* at 5-7.

[452] *Id.* at 8-11.

[453] *Id.* at 13.

[454] *Id.* at 13-15.

[455] *Id.* at 16.

that characterizing the component's microbial burden is essential for creating an adequate sterilization procedure.[456] Similarly, the Guidance offers recommendations on appropriate sterile preparation for containers and closures.[457]

### Endotoxin Control and Time Limitations

Contamination from endotoxin[458] is a particular concern for parenteral administration of sterile products, particularly in certain susceptible situations (e.g., neonate patients, large-volume doses).[459] FDA cautions that sterilizing-grade filters and moist heat sterilization have not been shown effective in controlling endotoxin, and as such, other sterilization techniques, such as high-temperature dry heat, are advised.[460] In addition, time limitations are suggested to avoid compromising sterilization filtration and other processes through accumulation of bio-burdens and endotoxins.[461]

### Process Validation

The Guidance also includes a discussion on qualification and validation studies. The recommendations include a detailed outline for appropriate process simulations in which a microbiological growth medium is exposed to product contact surfaces.[462] FDA recommends that process simulation closely replicate actual manufacturing conditions in order to serve as an effective process validation.[463] Other elements of these simulations discussed are the frequency, duration, size, and speed of the test runs as well as processing the growth medium, and interpretation of results.[464] Lastly, the process validation section provides suggestions to determine efficacy of filtration and equipment sterilization.[465]

### Laboratory Controls, Sterility Testing, and Batch Record Review

As aseptic processing is concerned with protection from microbial contamination, the Guidance stresses the importance of laboratory controls, and establishment of an environmental monitoring program.[466] The program should include a monitoring methodology, action levels and determination of disinfection efficacy.[467] In addition, the laboratory controls should characterize any microbiological hazards that are present during processing.[468]

The Guidance notes that adequate sterility testing is essential to detect potential contamination in a sterile product batch.[469] Elements of appropriate testing include microbiological laboratory controls, and sampling and incubation procedures.[470] In addition, a diligent investigation protocol for OOS results should be utilized.[471]

---

[456] STERILE DRUG PRODUCTS, *supra* note 446, at 16.

[457] *Id.* at 17-18.

[458] Endotoxin refers to "outer membrane of the cell wall of Gram-negative bacteria." KENNETH TODAR, Bacterial Endotoxin, in TODAR'S ONLINE TEXTBOOK OF BACTERIOLOGY (2008), http://textbookofbacteriology.net/endotoxin.html.

[459] STERILE DRUG PRODUCTS, *supra* note 446, at 19.

[460] *Id.*

[461] *Id.* at 20.

[462] *Id.* at 20-21.

[463] *Id.* at 21.

[464] *Id.* at 22-26.

[465] STERILE DRUG PRODUCTS, *supra* note 446, at 27-31.

[466] *Id.* at 32-33.

[467] *Id.* at 32-25.

[468] *Id.* at 35.

[469] *Id.* at 38.

[470] *Id.*

[471] STERILE DRUG PRODUCTS, *supra* note 446, at 39.

When available evidence cannot determine the cause of contamination, the impacted batches should be rejected as noncompliant.[472] Finally, the Guidance discusses appropriate documentation of batch records, with the inclusion of all in-process and laboratory control results, among other information.[473]

**Other Requirements**

Biologic products are subject to other requirements in addition to the CGMPs located in 21 C.F.R. parts 210 and 211. In particular, biologic products are subject to 21 C.F.R. parts 600 through 680 or 21 C.F.R. part 1271 for human cell, tissue, or cellular or tissue-based products (HCT/Ps). The CGMPs and biologic product requirements are intended to supplement each other, however, in the event of a conflict, the regulation more specifically applicable supersedes the more general regulation.[474]

Blood products, for example, have additional CGMPs located in 21 C.F.R. part 606. FDA issued a guidance document entitled, "Guideline for Quality Assurance in Blood Establishments," that provides recommendations on quality assurance procedures to be employed in conjunction with both general CGMPs and blood product CGMPs.[475] Particular emphasis is placed on the quality control/assurance unit responsibilities.[476] Given the complexities associated with the blood donation process, the Guideline provides detailed recommendations related to training and education that are more specific than recommendations for products regulated solely under parts 210 and 211.[477] For example, competency evaluations are to include observation of regulated tasks related to donor suitability, such as collection of health history information, and completion of the blood donation record.[478] While adverse event investigation and reporting is required for other drug products, the focus in those cases is the recipient of the product. The Guideline also contains recommendations related to adverse reactions experienced by blood donors.[479]

HCT/Ps are regulated as a device, drug, or biological product.[480] These products are subject to "current Good Tissue Practice" (cGTP) requirements.[481] If regulated as a device, HCT/Ps must abide by device quality system regulations.[482] If they are regulated as drugs or biologics, then the general CGMPs are applicable.[483] Core elements of the cGTP requirements include facilities, environmental control, equipment, supplies/reagents, recovery, processing controls, labeling controls, storage, distribution, and donor eligibility determinations.[484] Similar to the additional requirements for biologic products, the cGTPs are intended to supplement the CGMPs and in the event of a conflict, the more specific regulation prevails.[485]

---

[472] *Id.*

[473] *Id.* at 42.

[474] 21 C.F.R. § 210.2(a).

[475] See Food & Drug Admin., Guideline for Quality Assurance in Blood Establishments (1995), http://www.fda. gov/downloads/BiologicsBloodVaccines/GuidanceComplianceRegulatoryInformation/Guidances/Blood/ucm164981.pdf.

[476] *Id.* at 5-10.

[477] *Id.* at 7-9.

[478] *Id.*

[479] *Id.* at 10.

[480] Food & Drug Admin., Guidance for Industry: Current Good Tissue Practice (cGTP) and Additional Requirements for Manufacturers of Human Cells, Tissues, and Cellular and Tissue-Based Products (HCT/PS) 6 (2011), http:// www.fda.gov/downloads/biologicsbloodvaccines/Guidancecomplianceregulatoryinformation/Guidances/tissue/ucm285223.pdf [CGTP & HCT/Ps Guidance].

[481] 21 C.F.R. §§ 1271.145-1271.320.

[482] CGTP & HCT/Ps Guidance, *supra* note 480, at 6.

[483] *Id.*

[484] *Id.* at 3.

[485] 21 C.F.R. § 210.2(a).

## 8. CGMP Requirements for PET Drugs

Positron emission tomography (PET) drugs assist medical imaging through their radiopharmaceutical properties.[486] Historically, PET drugs were produced in small quantities by researchers and academics; however, PET drug production has grown with the increased use of medical imaging.[487] FDA was directed to issue CGMPs for PET drugs under the Food and Drug Administration Modernization Act of 1997.[488] The PET CGMPs are similar in form to the general CGMPs of 21 C.F.R. parts 210 and 211. Specifically, the PET CGMPs cover requirements relating to personnel and resources, quality assurance, facilities and equipment, components/containers/closures, production and process controls, laboratory controls, finished product controls, packaging/labeling, distribution, complaint handling, and records.[489] The PET CGMPs are codified in 21 C.F.R. part 212.

Given the unique nature and history of PET drugs there was some uncertainty as to the difference of PET drug production, and the practice of pharmacy.[490] FDA decided that production of PET drugs, which trigger the PET CGMPs, includes all activities up to the final release of the finished dosage form.[491] When a finished dosage PET drug is received by someone authorized to administer the PET drug, the subsequent usage or dispensing is considered practice of medicine and pharmacy, which is not subject to the PET CGMPs.[492] FDA also has another PET CGMP Guidance "PET Drugs – Current Good Manufacturing Practice (CGMP) (Small Entity Compliance Guide)," however the agency has stated this Guidance is substantively identical to its 2009 PET CGMP Guidance.[493]

## 9. CGMP Requirements for Combination Products

In January 2015, FDA issued "Current Good Manufacturing Practice Requirements for Combination Products" as draft guidance. This document provides an overview of FDA's final rule governing CGMP requirements for combination products, issued January 22, 2013. A combination product is described in 21 CFR part 3 as a product that contains "constituent parts" that are regulated separately as drugs, devices, or biological products.[494] The final rule clarifies that the regulatory status of each "constituent part" applies to the "constituent part" even after it is combined with other "constituent parts" and is required to comply with the applicable CGMP requirements for that part. For constituent parts that are manufactured separately and not co-packaged, the CGMP requirements for the constituent part do not change. For single-entity and co-packaged products, there are two options. The first option is to comply with the CGMP requirements for all constituent parts of the product. In other words, a drug-device combination product would be required to comply with all requirements in 21 CFR parts 210 and 211, plus all requirements in the quality system regulations (QSR), set forth at

---

[486] FOOD & DRUG ADMIN., GUIDANCE FOR INDUSTRY., GUIDANCE: PET DRUGS – CURRENT GOOD MANUFACTURING PRACTICE (CGMP) 4 (2009), http://www.fda.gov/downloads/Drugs/GuidanceComplianceRegulatoryInformation/Guidances/ucm070306.pdf [PET DRUGS GUIDANCE].

[487] *Id.* at 2.

[488] *Id.* at 1.

[489] See 21 C.F.R. pt. 212.

[490] PET DRUGS GUIDANCE, *supra* note 486, at 5.

[491] *Id.*

[492] *Id.*

[493] FOOD & DRUG ADMIN., GUIDANCE: GUIDANCE: FDA OVERSIGHT OF PET DRUG PRODUCTS QUESTIONS AND ANSWERS Q74 (2012), http://www.fda.gov/downloads/Drugs/GuidanceComplianceRegulatoryInformation/Guidances/UCM290024.pd.

[494] Specific combination products include: "A product comprised of two or more regulated components, i.e., drug/device, biologic/device, drug/biologic, or drug/device/biologic, that are physically, chemically, or otherwise combined or mixed and produced as a single entity;" or two such products packaged together or separately, but intended for use only in combination. 21 CFR §3.2(e).

21 CFR part 820. In order to avoid inefficiency, the final rule offers a "streamlined approach," which requires manufacturers to demonstrate full compliance with either the drug CGMPs or device requirements under the QSR. For example, a drug-based CGMP streamlining approach would require full compliance with 21 CFR parts 210 and 211, plus compliance with the following provisions from the QSR: management responsibility (820.20), design controls (820.30), purchasing controls (820.50), corrective and preventive action (820.100), installation (820.170), and servicing (820.200).[495]

The draft guidance provides a brief summary of CGMP provisions that are relevant to combination product streamlining approaches, and a discussion of the application of CGMP requirements to certain combination products (i.e. prefilled syringes, drug coated mesh, and drug eluting stents), in order to illustrate certain issues that might arise with respect to these products.

## 10. Compounding

In July 2014, FDA issued an Interim Draft Guidance, "Current Good Manufacturing Practice – Interim Guidance for Human Drug Compounding Outsourcing Facilities Under Section 503B of the FD&C Act" (the Interim Draft Guidance). The Interim Draft Guidance delineates the agency's expectations regarding outsourcing facility compliance with CGMP requirements in 21 CFR parts 210 and 211. Outsourcing facilities, as defined in 2013 Drug Quality and Security Act, are facilities engaged in the compounding of sterile drugs that have elected to register with FDA. The Interim Draft Guidance will be in effect for these facilities until more specific CGMP final regulations are promulgated.[496]

In the Interim Draft Guidance, FDA differentiates outsourcing facilities from conventional drug manufacturers by tailoring "CGMP requirements to the nature of the specific compounding operations conducted by the outsourcing facilities while maintaining the minimum standards necessary to protect patients from the risks of contaminated or otherwise substandard compounded drug products."[497] The Interim Draft Guidance focuses primarily on those operations that pose the highest risk to patient safety, including those aspects of 21 CFR Part

---

[495] 21 CFR §4.4(b)(1). Under 21 CFR §4.4(b)(2), a device streamlining approach would require full compliance with the QSR, plus compliance with 21 CFR 211.84, testing and approval or rejection of components, drug product containers, and closures; 21 CFR 211.103, calculation of yield, 21 CFR 211.132, tamper-evident packaging requirements for over-the-counter (OTC) human drug products, 21 CFR 211.137, expiration dating, 21 CFR 211.165, testing and release for distribution, 21 CFR 211.166, stability testing, 21 CFR 211.167, special testing requirements, and 21 CFR 211.170, reserve samples.

[496] FDA has published other guidance documents specific to outsourcing facilities. See FOOD & DRUG ADMIN., GUIDANCE FOR INDUSTRY: REPACKAGING OF CERTAIN HUMAN DRUG PRODUCTS BY PHARMACIES AND OUTSOURCING FACILITIES (2017), http://www.fda.gov/downloads/Drugs/GuidanceComplianceRegulatoryInformation/Guidances/UCM434174.pdf; FOOD & DRUG ADMIN., GUIDANCE FOR INDUSTRY: COMPOUNDING AND REPACKAGING OF RADIOPHARMACEUTICALS BY OUTSOURCING FACILITIES (2016), http://www.fda.gov/downloads/Drugs/GuidanceComplianceRegulatoryInformation/Guidances/UCM534812.pdf; FOOD & DRUG ADMIN., GUIDANCE FOR INDUSTRY: ELECTRONIC DRUG PRODUCT REPORTING FOR HUMAN DRUG COMPOUNDING OUTSOURCING FACILITIES UNDER SECTION 503B OF THE FEDERAL FOOD, DRUG, AND COSMETIC ACT (2016), http://www.fda.gov/downloads/Drugs/GuidanceComplianceRegulatoryInformation/Guidances/UCM424303.pdf; FOOD & DRUG ADMIN., GUIDANCE FOR INDUSTRY: ADVERSE EVENT REPORTING FOR OUTSOURCING FACILITIES UNDER SECTION 503B OF THE FEDERAL FOOD, DRUG, AND COSMETIC ACT (2015), http://www.fda.gov/downloads/Drugs/GuidanceComplianceRegulatoryInformation/Guidances/UCM434188.pdf; FOOD & DRUG ADMIN., GUIDANCE FOR ENTITIES CONSIDERING WHETHER TO REGISTER AS OUTSOURCING FACILITIES UNDER SECTION 503B OF THE FEDERAL FOOD, DRUG, AND COSMETIC ACT (2015), http://www.fda.gov/downloads/Drugs/GuidanceComplianceRegulatoryInformation/Guidances/UCM434171.pdf; FOOD & DRUG ADMIN., GUIDANCE FOR INDUSTRY: FEES FOR HUMAN DRUG COMPOUNDING OUTSOURCING FACILITIES UNDER SECTIONS 503B AND 744K OF THE FD&C ACT (2014), http://www.fda.gov/downloads/Drugs/GuidanceComplianceRegulatoryInformation/Guidances/UCM391102.pdf; FOOD & DRUG ADMIN., GUIDANCE FOR INDUSTRY: REGISTRATION OF HUMAN DRUG COMPOUNDING OUTSOURCING FACILITIES UNDER SECTION 503B OF THE FD&C ACT (2014), http://www.fda.gov/downloads/Drugs/GuidanceComplianceRegulatoryInformation/Guidances/UCM377051.pdf.

[497] Interim Draft Guidance at 2.

211 relating to sterility assurance and safety of compounded drugs, with respect to strength, and labeling or drug product mix-ups. FDA's inspectional and enforcement efforts will focus on these areas.

## a. CGMPs Applicable to Outsourcers

The Interim Draft Guidance outlines the agency's thinking regarding outsourcing facility obligations related to facility design, control systems and procedures for maintaining suitable facilities, environmental and personnel monitoring, equipment, containers and closures, components, production and process controls, release testing, laboratory controls, stability/expiration dating, packaging and labels, and quality assurance activities/complaint handling. FDA has requested public comments on alternative approaches that would reduce the need for laboratory testing of incoming components, and the need for facilities to have an in-house laboratory.

The Interim Draft Guidance identifies specific CGMP requirements under 21 CFR part 211 that are critical to ensuring the quality of compounded sterile drug products. It appears that these requirements are more extensive than compounders had anticipated, especially with respect to release and stability testing, and to the responsibilities of the quality control unit.

In certain instances, FDA intends to use its enforcement discretion if an outsourcing facility complies with criteria set forth in the Interim Draft Guidance. Some highlights include:

- Incoming Inspection: Outsourcing facilities are required to comply with requirements in 21 CFR 211.84(d) and 211.67(b) to examine each lot of equipment, containers, and closures prior to use, to verify identity and test to ensure that the equipment, container, or closure meets specifications.
    - Components generally must be re-tested or re-examined after lengthy storage or exposure to conditions that might adversely affect the component, but *not* if the component is stored under the supplier's labeled storage conditions.
    - FDA will exercise enforcement discretion regarding incoming inspection if the component is a finished drug product, purchased directly from the manufacturer, has had its label examined to verify that the component meets required specifications, and the package integrity has been verified. Any components not meeting acceptance criteria must be rejected.
    - Outsourcing facilities are also required to test bulk APIs and excipients to verify identity, and evaluate for conformity to specifications. Lot testing can be skipped if a Certificate of Analysis (COA) accompanies the lot, and meets the requirements in the Interim Draft Guidance.
    - FDA will exercise its enforcement discretion regarding testing of certain non-sterile components (e.g., water) purchased and certified as sterile and non-pyrogenic, and accompanied by a COA.
    - FDA requested comments on alternative approaches that might reduce the need for duplicative testing by multiple outsourcing facilities, while still giving them confidence in the quality of incoming components.
- Production and Process Controls:
    - Outsourcing facilities must document that key process parameters are controlled and deviations are justified.
        - Batch records must describe batch yield (actual v. calculated).
        - Outsourcing facilities must investigate any failure to meet acceptance criteria before approving lot release and a failure may result in rejection of the lot.

  - o Hold time assessments must be conducted as part of the process for validating sterility assurance.
    - For example, sterile drug products that are terminally sterilized should achieve a $10^{-6}$ sterility assurance level for proper validation of the sterilization process.
- Release Testing:
  - o The quality control unit must ensure that testing is conducted and test results demonstrate that the finished drug product meets specifications before batch release.
  - o FDA will exercise enforcement discretion for sterility testing if the batch consists of fewer than 10 dosage units compounded pursuant to a prescription for a single patient, so long as the unit is labeled with a beyond use date (BUD) that provides reasonable assurance of chemical and physical stability. Note that this is more restrictive than the standard established by the United States Pharmacopeia (USP) and adopted by many pharmacy compounders. For high-risk compounded sterile preparations, USP <797> requires release testing on batches of more than 25 dose units, multi-dose vials intended for multiple patients, or for preparations exceeding stated storage temperature conditions.[498]
  - o FDA also plans to exercise enforcement discretion for other testing for batches of less than 10 units (e.g., identity, strength, and particulate) if testing is conducted on every other batch or once for every 10 units.
- Stability Testing:
  - o Outsourcing facilities must establish stability programs to determine appropriate storage conditions and expiration dates, in accordance with 21 CFR § 211.166. In addition, outsourcing facilities must conduct stability testing to ensure that compounded products remain stable and sterile through the expiry period. FDA will exercise enforcement discretion regarding stability studies if the outsourcing facility establishes a BUD (1) in accordance with the Interim Draft Guidance, (2) literature or other scientific information support the chemical and physical stability of the BUD, and (3) the BUD is the expiration date.
- Packaging and Labeling: Labeling operations must have controls in place to prevent mix-ups and SOPs must be in place to ensure that the requirements of 21 CFR §§ 211.122, 211.125, 211.130 and 211.134 are met.
- Quality Assurance and Complaint Handling:
  - o The Interim Draft Guidance describes the roles and responsibilities of a quality control unit, which include:
    - Discrepancy and failure investigations (Note: the Interim Draft Guidance does not specifically require outsourcing facilities to retain samples, but it is not clear how an outsourcing facility would meet the requirement to conduct a thorough investigation without retaining samples).
    - Sampling and testing to ensure finished drug products meet release criteria.
    - Review of records to evaluate quality standards for each drug product.
    - Adverse event reporting.

The Interim Draft Guidance identifies a substantial portion of the CGMP regulations as "critical." Certain CGMPs are not specifically mentioned; however, their applicability is implied. For example, there

[498] USP <797> at 39–40.

is no citation to the CGMP requirement for adequate ventilation and air filtration under 21 CFR § 211.46, but air quality and cleanliness are discussed at length in the Interim Draft Guidance.[499] Similarly, there is no reference to yield calculation under 21 CFR § 211.103, yet the Interim Draft Guidance states a requirement to compare the "actual batch output (yield) . . . to the projected (calculated) output for each drug product."[500] Also, personnel training requirements are detailed under recommendations for aseptic drug processing.[501]

### b. Other CGMP Requirements Unclear

The Interim Draft Guidance appears to address every general CGMP category, except for Holding and Distribution (21 CFR §§ 211.142, 211.150) and Returned and Salvaged Drug Products (21 CFR §§ 211.204, 211.208).

## 11. Circumstances That Constitute Delaying, Denying, Limiting, or Refusing a Drug Inspection

The Food and Drug Administration Safety and Innovation Act (FDASIA) strengthened FDA's authority regarding inspection refusals by clarifying that delaying, denying, limiting or refusing to permit an inspection causes a drug to be adulterated.[502]

The "refusal to permit entry or inspection as authorized by section 704" has long been a prohibited act. 21 U.S.C. § 331(f). However, because a refusal now causes all drug products in that establishment to be adulterated, Congress required the agency to elaborate on circumstances that would fall within the scope of 21 U.S.C. § 351(j). FDA's Final Guidance entitled "Circumstances that Constitute Delaying, Denying, Limiting, or Refusing a Drug Inspection," issued in October 2014, provides FDA's current thinking regarding conduct that constitutes prohibited conduct.[503]

The Guidance provides examples of unacceptable delays related to scheduling, inspection management, and production of records. Examples of an inspection denial include (1) not allowing an FDA investigator to inspect the facility because certain staff members are not present; and (2) informing FDA that the facility does not engage in drug manufacturing operations.

---

[499] Interim Draft Guidance at 3-4.

[500] Interim Draft Guidance at 11.

[501] Interim Draft Guidance at 12.

[502] 21 U.S.C. § 351(j) (deeming a drug adulterated if "it has been manufactured, processed, packed, or held in any factory, warehouse, or establishment . . . [where the owner] delays, denies, or limits an inspection, or refuses to permit entry or inspection"). See Food & Drug Admin., *Warning Letter # 320-17-03 to Beijing Taiyang Pharmaceutical Industry Co Ltd* (October 19, 2016), https://www.fda.gov/iceci/enforcementactions/warningletters/2016/ucm527005.htm ("[O]ur investigators observed through a window a warehouse containing numerous drums bearing your company's label. When our investigators requested access to this warehouse, you barred them from entering the warehouse to examine the containers or the material in them without giving a reasonable explanation. The following day, you gave our investigators access to the warehouse. However, upon entry they observed that a significant number of drums [from the previous day] had been removed and were not available for inspection.")

[503] The Final Guidance also raises the question whether FDA could consider these types of conduct to be prohibited acts in inspections related to other types of FDA-regulated products.

Limiting an inspection could include a refusal to permit FDA investigators access to certain areas without a reasonable explanation. Examples include:

- ordering the discontinuation of manufacturing during the duration of the FDA inspection;
- limiting FDA's opportunity to perform direct observation; and
- "unreasonably restrict[ing] entry to a particular facility or a portion of the facility."

The Guidance also states that limiting the investigator's ability to take photographs without a reasonable explanation, providing the investigator documents that have been "unreasonably redacted," or refusing to allow FDA to collect samples constitutes a limitation of the inspection.

Finally, the Guidance provides examples of inspection refusals, which include active and passive conduct. In certain instances, FDA may not consider conduct described in the Guidance to be a prohibited act if the facility provides a "reasonable explanation" and the issue is resolved within a reasonable period of time.[504] Examples of reasonable conduct, as described in the Guidance, include requiring an investigator to comply with safety requirements or gowning procedures prior to inspection of a particular area.[505] In addition, if a facility's manufacturing schedule does not include production of a particular drug during FDA's requested time period[506] or if key personnel are not initially available for an unannounced inspection, FDA could consider these limitations to be reasonable.[507] In terms of an inability to provide immediate access to records, a potentially reasonable explanation might be that the records are currently in use for a manufacturing process or that the records need to be translated to meet the investigators request for English translations. While there is no bright line test for what is "reasonable," if a delay occurs during the course of an FDA inspection, the manufacturer must provide a factual, truthful explanation for the delay, and must ensure that the delay is resolved within a reasonable amount of time.

The Final Guidance highlights the need for proper inspection management training prior to an inspection. Companies need to understand the scope of FDA's inspection authority, and the legal significance of an inspection refusal. In addition, companies need to establish well-defined procedures for FDA's entry and notice of inspection, and need to understand what to do if those procedures are not followed. For example, law enforcement personnel who are conducting a criminal investigation will not provide a Form FDA-482, Notice of Inspection. It is important for outside counsel to be involved in order to ensure that the client manages these investigations appropriately.

Outside counsel should assist the company in preparing employees for interactions with FDA investigators. For example, employees must be advised of the need to be truthful in all cases, and the legal risks of not

[504] See, e.g. GUIDANCE ON DELAYING, DENYING, LIMITING, OR REFUSING INSPECTIONS, *supra* note 503, at 5 ("In instances where the facility provides a reasonable explanation for delaying production of records, the facility should also ensure that the resulting delay is of a reasonable duration.").

[505] *See id.* at 7 ("Training specified by the Occupational Safety and Health Administration is required before an individual may enter a particular area of the facility, and the FDA investigator has not completed such training."); *id.* at 5 ("A facility does not provide the FDA investigator access to aseptic processing areas until the investigator accommodates the facility's documented gowning procedures.").

[506] *Id.* at 4 ("Manufacturing at the facility is not on-going, for example running only one manufacturing campaign per month . . . ").

[507] *Id.* at 6 ("At the beginning of an unannounced inspection, appropriate personnel are not immediately available to accurately answer the FDA investigator's questions."). Comparatively, preventing an inspection from occurring due to the absence of certain personnel may be considered a denial of the inspection. *Id.* ("A facility does not allow the FDA investigator to inspect the facility because certain staff members are not present, without a reasonable explanation.").

doing so. Similarly, with respect to documents, employees must be advised not to alter or hide documents, and the legal consequences of not providing requested information.

## 12. Data Integrity

Product quality and compliance with the FDCA and all applicable regulatory requirements are, at the most basic level, supported by data. The integrity of data collected, recorded, and maintained by a facility is critical to ensuring that only high quality and safe drugs are manufactured. These data are relied upon not only by drug product manufacturers, but by FDA and other government regulators, healthcare professionals, and patients. A failure to ensure data integrity can have profound consequences, ranging from contaminated or unsafe products entering the marketplace to serious injury, and even death. For this reason, the FDCA establishes requirements to ensure that drug products are safe and effective for their intended uses. These requirements are further described in FDA's "Application Integrity Policy,"[508] (the AIP), CGMP regulations, and most recently, FDA's draft guidance, "Data Integrity and Compliance with CGMP: Guidance for Industry," [509] (DI Draft Guidance).

The AIP describes FDA's general approach with respect to representations and data contained in new drug applications. Submitting data that lacks integrity, such as data that is untrue, false, inaccurate, or lacks a source in an actual event, is a wrongful act, and can result in denial or withdrawal of an application, and a possible enforcement action. Under the AIP, FDA must document a "pattern or practice" of wrongful conduct that raises significant questions about the reliability of data submitted to support an application(s) and it must make a determination that the wrongful act is material to the agency's decisionmaking process regarding approval.[510]

Applications should not be approved if the data are unreliable, or when there are data integrity, safety or efficacy concerns, until FDA is satisfied that the issues in question are resolved. AIP outlines recommended corrective actions by which applicants may seek to restore FDA's confidence in their integrity and permit the agency to proceed with substantive scientific review of their premarket applications. These include:

- Full cooperation with FDA investigations to determine the cause and scope of any wrongful acts and to assess the effects of the acts on the safety, effectiveness, or quality of products;
- Full identification of all individuals potentially associated with or involved in the wrongful acts to ensure that they are removed from any substantive authority on matters under the jurisdiction of FDA;
- A credible internal review by an outside consultant—intended to supplement FDA's ongoing, comprehensive review—designed to identify all instances of wrongful acts associated with applications submitted to FDA, including any discrepancies between manufacturing conditions identified in approved applications and manufacturing conditions during actual production; and
- A commitment, in writing, to developing and implementing a corrective action operating plan to assure the safety, effectiveness, and quality of their products.

In 2014, FDA issued a number of Warning Letters with data integrity observations. These Warning Letters identified conduct such as audit trail functions that were disabled at time of inspection; "unofficial" testing of

---

[508] Food & Drug Admin., Application Integrity Policy (2004), https://www.fda.gov/downloads/ICECI/EnforcementActions/ApplicationIntegrityPolicy/ucm072631.pdf [FDA AIP]

[509] Food & Drug Admin., Data Integrity and Compliance with CGMP: Guidance for Industry (2016), https://www.fda.gov/downloads/Drugs/GuidanceComplianceRegulatoryInformation/Guidances/UCM495891.pdf [DI Draft Guidance].

[510] FDA AIP, *supra* note 508, at 10.

samples, discarded results, and results from additional tests reported as results of record; discarded hard copy records found during inspection; testing into compliance; failing results not included in official lab control records, and not reported or investigated; batch records signed by individuals who did not perform the review; and activities that were not performed, but were recorded in batch records.[511]

These types of findings prompted FDA to issue the DI Draft Guidance in April 2016, to clarify the role of data integrity in the drug manufacturing environment. The DI Draft Guidance applies to finished pharmaceuticals, regulated under 21 CFR parts 210 and 211, and positron emission tomography (PET) drugs, regulated under part 212.

The DI Draft Guidance highlights certain regulatory requirements for data integrity, such as requirements for data security[512] and preservation,[513] concurrent documentation,[514] control of laboratory records,[515] retention of originals, "true copies," and "accurate reproductions" of records,[516] and requirements to preserve complete records of all data and tests performed.[517]

The DI Draft Guidance is structured as a series of questions and answers, and defines "data integrity" as "the completeness, consistency, and accuracy of data. Complete, consistent, and accurate data should be

---

[511] Based on a search of FDA's Warning Letter Database, the number of CGMP-related FDA Warning Letters that include data integrity findings has continued to increase since 2014: five issued in 2014, seven issued in 2015, and 11 issued in 2016. *See generally* Food & Drug Admin, *Inspections, Compliance, Enforcement, and Criminal Investigations: Warning Letters*, https://www.fda.gov/ICECI/EnforcementActions/WarningLetters/default.htm (Last Updated February 21, 2017). The inspections causing the warning letters continue to identify particularly concerning issues. *See, e.g.,* Food & Drug Admin., *Warning Letter # 320-16-29 to Pan Drugs Limited* (August 25, 2016), https://www.fda.gov/iceci/enforcementactions/warningletters/2016/ucm518540.htm ("[Y]our quality unit failed to identify data integrity issues in 11 batch production records reviewed by our investigator. Your production manager admitted that he falsified the signatures of other employees in the 'Prepared By,' 'Reviewed By,' 'Approved By,' and 'Authorized By' sections."); Food & Drug Admin., *Warning Letter # 320-16-19 to Chongqing Lummy Pharmaceutical Co., Ltd.* (June 21, 2016), https://www.fda.gov/iceci/enforcementactions/warningletters/2016/ucm508291.htm ("[O]n March 4, 2016, your analyst set the GC personal computer (PC) clock back to make it appear as if testing had been done seven months earlier – on August 3, 2015 . . ."); Food & Drug Admin., *Warning Letter # 320-16-10 to Polydrug Laboratories Pvt. Ltd.* (April 14, 2016), https://www.fda.gov/iceci/enforcementactions/warningletters/2016/ucm496623.htm ("During the inspection our investigator found a torn sheet of paper titled 'Product Quality Complaints' on the floor of your warehouse. We compared it to your firm's official complaint log and discovered that only 2 of the 17 customer complaints on the torn sheet were recorded in your firm's official complaint log."); Food & Drug Admin., *Warning Letter # 320-15-12 to Mahendra Chemicals* (July 13, 2015), https://www.fda.gov/iceci/enforcementactions/warningletters/2015/ucm455345.htm ("[O]ur investigators found backdated batch production records dated February 10 to February 25, 2014, signed by your Production Manager and Technical Director in the 'Batch Manufacturing Record Reviwed [sic] by' section. The Technical Director stated that he was not in the facility on these dates and was 'countersigning' for another person who allegedly performed these review activities. However, these records did not contain signatures (contemporaneous or otherwise) of the alternate reviewer who purportedly conducted the review."); Food & Drug Admin., *Warning Letter # 320-14-13 to Marck Biosciences Ltd.* (July 8, 2014), https://www.fda.gov/iceci/enforcementactions/warningletters/2014/ucm409898.htm ("Employees interviewed during the inspection admitted that your firm recorded activities in batch records that were not performed.").

[512] 21 C.F.R. § 211.68(b) ("Appropriate controls shall be exercised over computer or related systems to assure that changes in master production and control records or other records are instituted only by authorized personnel.").

[513] 21 C.F.R. § 211.110(b).

[514] 21 C.F.R. § 211.100(b) ("Written production and process control procedures shall be followed in the execution of the various production and process control functions and shall be documented at the time of performance.").

[515] 21 C.F.R. § 211.160(a) ("The requirements in this subpart shall be followed and shall be documented at the time of performance. Any deviation from the written specifications, standards, sampling plans, test procedures, or other laboratory control mechanisms shall be recorded and justified.").

[516] 21 C.F.R. § 211.180(d) ("Records required under this part may be retained either as original records or as true copies such as photocopies, microfilm, microfiche, or other accurate reproductions of the original records.").

[517] 21 C.F.R. § 211.188 ("Batch production and control records shall be prepared for each batch of drug product produced and shall include complete information relating to the production and control of each batch"), § 211.194(a) ("Laboratory records shall include complete data derived from all tests necessary to assure compliance with established specifications and standards . . .") and § 212.60(g) ("Each laboratory performing tests related to the production of a PET drug must keep complete records of all tests performed to ensure compliance with established specifications and standards . . . ").

attributable, legible, contemporaneously recorded, original or true copy, and accurate (ALCOA)."[518] The DI Draft Guidance is, for the most part, focused on electronic data, and provides FDA's current thinking regarding minimum requirements for audit trails, computer system validation, access controls, such as restricting rights to alter files or settings to personnel who are independent from those responsible for content of the records, and restrictions on the use of credentials. The DI Draft Guidance also provides recommendations consistent with those outlined in the AIP for addressing data integrity problems. Specifically, the DI Draft Guidance recommends hiring an independent auditor, determining the scope of data integrity lapses, and developing and implementing a global plan to correct data integrity problems, and removing personnel who have engaged in objectionable conduct with respect to data integrity from regulated manufacturing activities.

## 13. Contract Manufacturing

In November 2016, FDA issued "Contract Manufacturing Arrangements for Drugs: Quality Agreements: Guidance for Industry," (Quality Agreements Guidance) to provide guidance regarding the relationships between entities that introduce a drug into interstate commerce (the "Owner") and third parties that perform some or all of the manufacturing operations for the product (the "Contracted Facilities").[519] The guidance describes how Owners and Contracted Facilities can use Quality Agreements to define responsibilities associated with manufacturing activities in a manner that is consistent with CGMP requirements.

The Quality Agreements Guidance defines a quality agreement as "a comprehensive written agreement between parties involved in the contract manufacturing of drugs that defines and establishes each party's manufacturing activities in terms of how each will comply with CGMP."[520] A quality agreement should not be used to cover commercial terms, "such as confidentiality, pricing or cost issues, delivery terms, or limits on liability or damages."[521] In addition, while the guidance makes clear that CGMP regulations do not require quality agreements, it reminds industry that quality unit responsibilities and procedures are required to be documented in writing and that "[i]mplementing a written quality agreement can facilitate compliance with CGMP and, in particular, with 21 CFR 211.22(d)."[522] Most importantly, the guidance emphasizes that parties cannot, by agreement, shift CGMP responsibilities, whether established by statute or regulation. For example, with respect to batch release, the guidance makes clear that, under 21 C.F.R. § 211.22(a), owners are legally responsible for approving or rejecting product for release into interstate commerce and cannot delegate this responsibility to a Contracted Facility.

The guidance outlines the basic elements of a quality agreement as (1) the purpose and scope of the agreement; (2) definitions; (3) how to resolve disagreements; (4) responsibilities for specific manufacturing activities, such as how the parties' quality units will work together; and (5) revision and termination of the quality agreement. Of these elements, the guidance emphasizes that responsibilities for manufacturing activities, generally, are the most important to CGMP compliance, and that quality and change control of manufacturing activities, specifically, are "critical."[523]

---

518 DI Draft Guidance, *supra* note 509, at 2.

519 Food & Drug Admin., Contract Manufacturing Arrangements for Drugs: Quality Agreements: Guidance for Industry (2016), https://www.fda.gov/ucm/groups/fdagov-public/@fdagov-drugs-gen/documents/document/ucm353925.pdf [Quality Agreements Guidance].

520 *Id.* at 6.

521 *Id.*

522 *Id.* at 5.

523 *Id.* at 7.

The quality and change control aspects of manufacturing activities that should be covered by the quality agreement include the Contracted Facilities quality unit, as well facilities maintenance, materials management, product-specific quality considerations, and laboratory controls. The following are the types of questions the quality agreement should answer:

- Between the quality units for the Owner and the Contracted Facility, who is responsible for which quality activities?
- What is the role of each quality unit in audits and inspections, including responding to observations and determining corrective actions?
- Who will be maintaining, validating, and replacing the facilities and equipment used for the Owner's products?
- For component materials, who establishes specifications for testing, acceptance, storage, and transportation?
- What are the product-specific specifications, manufacturing details (e.g., batch numbering), expiration requirements, and process validation?
- How will learnings about the manufacturing of the product be shared between the Contracted Facility and the Owner, and vice versa?
- What is the process for establishing and conducting sampling and testing?
- Who will maintain original records?
- What are the processes for changing any of the above, including which changes need pre-approval by the Owner?

In addition to the questions above, communication around manufacturing activities is also emphasized throughout the Guidance. When drafting the communications portions of the quality agreement, the owner should consider the form (e.g., reports or summary tables, email or physical mail), triggers (e.g., unexpected events versus manufacturing deviations), regularity (e.g., weekly or daily), and timing (e.g., within 24 hours, the first Tuesday of each month). These requirements will also have to take into account whether the owner wishes to be made aware of issues related to the CMGP-regulated activities the Contracted Facility is performing for other Owners.

The Guidance also provides case scenarios that highlight situations frequently encountered by FDA investigators during inspections, such as responsibility for maintenance and upgrade of facilities, accuracy of batch records, responsibilities of contract laboratories, and responsibilities for change control.

## 14. Quality Metrics

As discussed above, in 2008, FDA determined that batches of heparin API manufactured in China were contaminated with oversulfated chondroitin sulfate (OSCS). The agency concluded that the contamination of crude heparin was intentional, to reduce costs of production. The heparin crisis highlighted other limitations on FDA's ability to ensure product quality.[524] On July 9, 2012, Congress granted FDA additional statutory

[524] Heparin: A Wake-Up Call on Risks to the U.S. Drug Supply, Issue Brief, *Pew Charitable Trust* (May 16, 2012), http://www.pewtrusts.org/en/research-and-analysis/issue-briefs/2012/05/16/heparin-a-wakeup-call-on-risks-to-the-us-drug-supply.

authority through FDASIA,[525] with several provisions related to the drug supply chain and implicating the quality metrics program. FDA relies on two provisions in particular, sections 705 and 706, as the statutory authority for FDA's proposed quality metrics program.[526]

Section 705 of FDASIA replaced the previous biennial inspection frequency for registered drug firms with a risk-based inspection program.[527] The risk-based model utilizes numerous factors to prioritize facility inspections including:

- The compliance history of the drug firm;
- The record, history, and nature of recalls linked to the drug firm;
- The inherent risk posed by the drug manufactured, prepared, propagated, compounded, or processed at the drug firm;
- The inspection frequency and history of the drug firm, including whether the firm has been inspected within the last four years;
- Whether the drug firm has been inspected by a foreign government; and
- Any other criteria deemed necessary and appropriate for purposes of allocating inspection resources.[528]

Section 706 of FDASIA allows FDA to request records in advance, or in lieu of, an establishment inspection.[529] These two provisions operate in concert: FDA will use quality metrics (i.e. records) collected under Section 706 to determine the risk profile of a facility for purposes of establishment inspection under the Section 705 risk-based model.

Sections 707 and 711 of FDASIA also influence the architecture of the quality metrics program. Section 707 allows FDA to deem a facility's drug products adulterated if the facility delays, denies, or limits an establishment inspection. FDA has indicated that the failure to provide quality metrics data will be viewed by FDA as a violation of Section 707.[530]

Section 711 of FDASIA amended section 501 of the FDCA by adding the following language:

> For purposes of paragraph [501](a)(2)(B), the term "current good manufacturing practice" includes the implementation of oversight and controls over the manufacture of drugs to ensure quality, including managing the risk of and establishing the safety of raw materials, materials used in the manufacturing of drugs, and finished drug products.[531]

Section 711 authorizes FDA to require finished product manufacturers to establish oversight and controls over its suppliers, including contract manufacturing organizations (CMOs). If a finished product manufactur-

---

[525] The Food and Drug Administration Safety and Innovation Act (FDASIA), Pub. L. No. 112-144, 126 Stat. 993. FDASIA also authorized the collection of FDA user fees for innovator drugs, medical devices, generic drugs, and biosimilar biologics. *Id.*

[526] Russell Wesdyk, *Quality Metrics Update: Stakeholder Feedback, Goals, and Gaps* (May 28, 2014) at slide 3, https://www.fda.gov/downloads/drugs/ucm401496.pdf. [Wesdyk, *Quality Metrics Update: Stakeholder Feedback…*].

[527] FDASIA, Pub. L. No. 112-144, 126 Stat. 993 (2012).

[528] *Id.*

[529] *Id.* § 706.

[530] U.S. Food & Drug Admin., *CDER's Quality Metrics Initiative: Program Update* at Slide 13, http://www.pqri.org/workshops/EVP2014/Mullin.pdf.

[531] 21 U.S.C. § 351.

er (or application holder) fails to establish oversight and controls related to raw materials, components, and contract manufactured finished products, the manufacturer's distributed products are not in compliance with CGMPs and will be deemed adulterated. Thereby, the introduction of those products in interstate commerce is a prohibited act. Arguably, section 711 could impose oversight responsibility on finished product manufacturers to ensure that CMOs and raw material suppliers report quality metrics, because FDA could claim a violation of Section 711against the finished product manufacturer if a supplier of raw materials or a contract manufacturer fails to submit quality metrics data to FDA.

Title X of FDASIA also equipped FDA with new authority to combat drug shortages. In particular, FDASIA imposed new requirements on manufacturers to provide early notification to FDA of potential drug shortages or other disruptions in the supply of a product.[532] FDA has stated that the quality metrics program will help prevent drug shortages, which will help FDA fulfill this requirement.

## 15. Establishment of FDA Office of Pharmaceutical Quality

As part of FDA's broad efforts to address drug quality, the agency established an Office of Pharmaceutical Quality (OPQ).[533] OPQ's mission is to ensure the American public has access to "quality medicines" with the goal of being "a global benchmark for the regulation of pharmaceutical quality."[534] By establishing consistent quality standards and clear expectations for industry, OPQ's goal is to centralize quality drug review by "integrating quality review, quality evaluation, and inspection" across the product life cycle.[535] Within OPQ is an Office of Surveillance, which will monitor, assess, and report on quality issues.[536] As part of these surveillance activities, the Office of Surveillance plans to use standardized quality metrics obtained from manufacturers to manage FDA's quality surveillance, inspection, and analysis program.[537]

To support FDA's risk-based inspection model, the goals of OPQ, and to help predict possible drug shortages, FDA reached out to industry in 2013 to explore the broader use of establishing quality metrics for manufacturing drugs.[538] On February 12, 2013, FDA requested comments on several topics to address drug shortages, including "the broader use of manufacturing quality metrics."[539] FDA posed several questions regarding quality metrics:

---

[532] FDASIA, Pub. L. No. 112-144, §§ 710, 712, 126 Stat. 993, 1070-72.

[533] Brian Hasselbalch, *FDA Drug Quality Regulation* (June 2014), https://www.fda.gov/downloads/Drugs/DevelopmentApprovalProcess/SmallBusinessAssistance/UCM407992.pdf. [Hasselbalch, FDA Drug Quality Regulation]

[534] Robert Iser, U.S. Food & Drug Admin., Office of Pharmaceutical Quality Presentation, *Global Drug Development and its Impact on CDER's Drug Review Process Symposium* (June 24, 2014).

[535] *Id.*

[536] Hasselbach, *FDA Drug Quality Regulation*, *supra* note 533, at 22.

[537] *Id.*

[538] Parenteral Drug Association Points to Consider: Pharmaceutical Quality Metrics at 1.1, http://journal.pda.org/content/68/5/535.full.pdf. [PDA Points to Consider]

[539] FDA Drug Shortages Task Force and Strategic Plan; Request for Comments, 78 Fed. Reg. 9928, Dkt. No. FDA-02013-N-0124 (Feb. 12, 2013).

- What metrics do manufacturers currently use to monitor production quality?
- To what extent do purchasers and prescribers use information about manufacturing quality when deciding how to purchase or utilize products?
- What kinds of manufacturing quality metrics might be valuable for purchasers and prescribers when determining which manufacturers to purchase from or which manufacturers' products to prescribe?
- What kinds of manufacturing quality metrics might be valuable for manufacturers when choosing a contract manufacturer?
- How frequently would such metrics need to be updated to be meaningful?[540]

These questions, and quality metrics broadly, were discussed at various conferences in 2013[541] and 2014.[542] In the fall of 2014, the program was expanded to include biologics regulated by CBER. Devices and combination products that are regulated by CDRH will not be covered by the quality metrics program.

In March 2014, the International Society for Pharmaceutical Engineering (ISPE) announced that it would conduct the industry's first Quality Metrics Pilot Program to refine metric selection and definitions and to develop standard metrics reporting to FDA.[543] McKinsey & Company, an independent third-party and global consulting firm, collected and analyzed data from participating companies. The pilot program, which began in summer 2014, tested a series of leading and lagging indicators based on the input received from both FDA and industry to standardize a set of metric definitions.[544] Additionally, the pilot program tested the feasibility of data collection across different companies.[545] For the pilot program, ISPE included the following metrics:

- "Industry consensus" metrics (lot acceptance rate, total complaints rate, confirmed OOS rate and U.S. recall events);
- Quantitative metrics (critical complaints rate, stability failure rate, unconfirmed OOS rate, right first time (rework/reprocessing) rate, APQR reviews completed on time, recurring deviations rate, and CAPA effectiveness rate);
- Technology-specific metrics (media fill and environmental monitoring); and
- Survey-based metrics (process capability and quality culture).[546]

Upon completion of the pilot, McKinsey provided two deliverables: 1) blinded industry-level analyses that was shared with ISPE and all participants; and 2) company-specific benchmarking reports showing where each site stands compared to other participants.[547]

---

540 *Id.*

541 Wesdyk, *Quality Metrics Update: Stakeholder Feedback…, supra* note 526, at 4.

542 *Id.* at 17.

543 ISPE News, "*ISPE to Conduct Quality Metrics Pilot Program*," (Mar. 12, 2014), http://www.ispe.org/news/2014/ispe-conduct-quality-metrics-pilot.

544 *Id.*

545 *Id.*

546 *Id.*

547 *Id.*

Following on the ISPE pilot, FDA issued a Draft Guidance, "Request for Quality Metrics." in July 2015.[548] FDA cited its new authority under FDASIA to prevent drug shortages as partial justification for the quality metrics program, stating that quality issues are one of the main causes of drug shortages.[549]

The Guidance proposed a mandatory quality metrics program, under which each covered establishment would be required to report ten discrete metrics data points on a quarterly basis, for a total of forty data points per covered establishment per year. FDA stated that it intended to request one report for each Finished Dosage Form (FDF) and one for each API, under the belief that proper oversight of third parties would mean that "one establishment will already possess or have access to all of the quality metrics data needed to submit such reports."[550] Below were the primary metrics:

- "Product Quality Complaint Rate" - the number of product quality complaints received for the product divided by the total number of lots of the product released in the same timeframe.
- "Lot Acceptance Rate" - 1 – X, where "X" is the number of specification-related rejected lots in a timeframe divided by the number of lots attempted by the same establishment in the same timeframe.
- "Invalidated Out-of-Specification (OOS) Rate" - the number of OOS test results for the finished product invalidated by the establishment divided by the total number of OOS test results divided by the total number of tests performed by the establishment in the same timeframe.
- "Annual Product Review" (APR) - the number of APRs or PQRs completed within 30 days of annual due date at the establishment divided by the number of products produced at the establishment.

FDA also requested comments on a number of optional metrics, including Senior Management Engagement, CAPA Effectiveness, and Process Capability/Performance.[551]

In November 2016, FDA issued a revision to its 2015 Guidance, "Submission of Quality Metrics Data."[552] The 2016 Guidance indicated that the program would be voluntary in nature, and made other adjustments

---

[548] Food & Drug Admin., Request for Quality Metrics: Guidance for Industry 1 (2015), https://docplayer.net/5964542-Request-for-quality-metrics-guidance-for-industry.html.

[549] *Id.* at 4-5 ("The agency has found that the majority of drug shortages stem from quality concerns – substandard manufacturing facilities or processes are discovered, or significant quality defects are identified in finished product, necessitating remediation efforts to fix the issue, which in turn, may interrupt production, and cause a shortage of drugs .... FDA intends to use these quality metrics, in part, as a tool ... that could potentially be predictive of drug supply disruption.")

[550] Presentation, Janet Woodcock, Karthik Iyer, & Alex Viehmann, Draft Guidance for Industry: Request for Quality Metrics (July 2015), https://www.fda.gov/downloads/Drugs/DevelopmentApprovalProcess/SmallBusinessAssistance/UCM456211.pdf. Prior to the Public Meeting on the Quality Metrics Program, which FDA convened on August 24, 2015, FDA sought comments on several questions, including: "Are the metrics requested from each business segment/type clear and appropriate?" Presumably FDA has chosen to draw a single distinction between business types, i.e., reporting establishments and covered establishments. As we discuss in our comments, it is not clear who is the responsible establishment with respect to API, and this needs to be clarified. Moreover, FDA has not acknowledged or discussed the fact that there may be differences depending on the types of products manufactured (e.g., injectable v. solid dose or prescription v. OTC) or the size of the company.

[551] Food & Drug Admin., Request for Quality Metrics: Guidance for Industry 12-13 (2015), http://docplayer.net/5964542-Request-for-quality-metrics-guidance-for-industry.html ("Was each APR or PQR reviewed and approved by the following: (1) the head of the quality unit, (2) the head of the operations unit; (3) both; or (4) neither? . . . What percentage of your corrective actions involved re-training of personnel (i.e., a root cause of the deviation is lack of training)? . . . whether the establishment's management calculated a process capability or performance index for each critical quality attribute (CQA) as part of that product's APR or PQR . . . whether the establishment's management has a policy of requiring a corrective action or preventive action (CAPA) at some lower process capability or performance index . . .").

[552] Food & Drug Admin., Submission of Quality Metrics Data: Guidance for Industry (2016), https://www.fda.gov/downloads/drugs/guidances/ucm455957.pdf.

to the proposed program, such as removing one of the metrics, the APR review, but increasing the number of requested data elements. Industry has questioned whether the program proposed by the 2016 Guidance would be voluntary, in light of the fact that FDA intends to "include the reporting of quality metrics as a factor in our surveillance inspection risk-based model [and] publish a list of reporters who provide a certain amount of information."[553] Industry has also noted that the program, as envisioned in the 2016 Guidance, would be quite burdensome, both for industry and FDA.

[553] Submission of Quality Metrics Data; Draft Guidance for Industry; Availability; Request for Comments 81 Fed. Reg. 85226, 85229 (Nov. 25, 2016).

# Chapter II
# How Does FDA Evaluate CGMP Compliance?

## A. Establishment Inspections

FDA is authorized to conduct inspections of "any factory, warehouse, or establishment in which . . . . drugs . . . are manufactured, processed, packed or held for introduction into interstate commerce or after such introduction."[1] FDA's inspection authority is quite broad, and FDA investigators are entitled to request and review "records, files, papers, processes, controls and facilities" bearing on whether drugs are adulterated, misbranded, or otherwise violative.[2] FDA, however, may not inspect:

- Financial data;
- Sales data other than shipment data;
- Pricing data
- Personnel data other than information establishing the qualifications of technical and professional personnel; and
- Certain research data except to the extent such information may be required to be made available for inspection or submitted to FDA for particular products.

Failure to comply with an inspection request authorized under section 704 is considered an inspection "refusal" and is a prohibited act under section 301(e)-(f) of the FDCA.[3] In addition, refusing an inspection renders adulterated a drug manufactured, processed, packed, or held in the facility for which the inspection was refused.[4]

In certain circumstances, FDA will conduct a pre-approval inspection (PAI) before a firm is allowed to introduce a product subject to a marketing application into interstate commerce. In conducting the PAI, FDA's objectives are to determine compliance with CGMPs, conformance with application commitments, authenticity and accuracy of data in the application, and adequacy of the firm's analytical methodology.[5]

---

[1] Federal Food, Drug, and Cosmetic Act § 704, 21 U.S.C. § 374(a)(1).

[2] Note that compounding pharmacies that are not registered as "outsourcing facilities" are exempt from FDA's records inspection authority. See *id.* § 374(a)(2)(A). Specifically, the FDCA states that FDA's records inspection authority does not extend to "pharmacies which maintain establishments in conformance with any applicable local laws regulating the practice of pharmacy and medicine and which are regularly engaged in dispensing prescription drugs or devices, upon prescriptions of practitioners licensed to administer such drugs or devices to patients under the care of such practitioners in the course of their professional practice, and which do not, either through a subsidiary or otherwise, manufacture, prepare, propagate, compound, or process drugs or devices for sale other than in the regular course of their business of dispensing or selling drugs or devices at retail." *Id.* Registered outsourcing facilities are subject to FDA's records inspection authority. *See* 21 U.S.C. § 353b(b)(4)(A).

[3] *Id.* § 331(e)-(f).

[4] 21 U.S.C. § 351(j).

[5] *See* Food & Drug Admin., Program No. 7346.832 Pre-Approval Inspections/Investigations, in Compliance Program Guidance Manual (2010), http://www.fda.gov/downloads/Drugs/DevelopmentApprovalProcess/Manufacturing/ QuestionsandAnswersonCurrentGoodManufacturingPracticesCGMPforDrugs/ucm071871.pdf.

FDA Compliance Program (cPGM) 7346.832 incorporates a risk-based approach to PAIs consistent with the CGMPs for the 21st Century Initiative and establishes two general categories for inspection: priority and discretionary. Priority PAIs are required if there is uncertainty about a manufacturer's ability to manufacture the product in compliance with CGMPs. For example, if the manufacturer is listed for the first time in an application or the applicant has never filed an application a PAI is required. Applications for products that are more difficult to manufacture or have significant changes to the manufacturing process, dosage form, or intended use are also priority PAIs.[6]

In evaluating whether a firm is in compliance with CGMPs, FDA investigators focus on identifying deficiencies that could lead to the manufacture or distribution of adulterated and/or misbranded products.[7] Investigators will confirm that a firm's registration and drug listing are in compliance[8] and will ensure that firms are reporting field alerts for NDA and ANDA products, as required by 21 C.F.R. § 314.81, and Biological Product Deviation Reports (BPDRs) for therapeutic biologics, as required under 21 C.F.R. § 600.14. Investigators may also inspect to determine whether the firm is in compliance with reporting requirements for adverse drug experiences.[9] The depth of an establishment inspection depends upon the reason FDA is conducting the inspection and the type of information FDA seeks, and whether there are suspected violations.[10]

As part of their preparation for an inspection, FDA investigators are instructed to review Establishment Inspection Reports (EIRs), which provide comprehensive discussions of prior inspections. Investigators also review applications (NDAs, ANDAs, BLAs, and INDs) for application products, and OTC monographs for non-application products. In order to identify potential or emerging problems with product quality, investigators review complaints, recalls, Field Alert Reports (FARs), and BPDRs. Investigators are also instructed to use this information to identify drugs that violate the FDCA due to misbranding, lack of approval, fraud, withdrawn applications or products that have been removed from the market because of safety concerns.[11]

FDA's Compliance Program Guidance Manual CP 7356.002 recommends a systems and risk based approach to establish inspections and identifies the following six systems for inspection: Facilities and Equipment, Materials, Laboratory Controls, Production, Packaging and Labeling and, most importantly, Quality. In a full inspection, FDA will cover at least four systems and two systems in an abbreviated inspection. The Investigations Operations Manual (IOM), which provides guidance to FDA's field investigators, emphasizes that "[d]uring the evaluation of the Quality System it is important to determine if top management makes sci-

---

6 *Id.* at 3.

7 Food & Drug Admin., Investigations Operations Manual 290 (2017), http://www.fda.gov/iceci/inspections/iom/default.htm [Investigations Operations Manual].

8 Any person who owns or operates an establishment that is "engage[d] in the manufacture, preparation, propagation, compounding, or processing of a drug or drugs" is required to register as an establishment and list drug products that the firm distributes. 21 C.F.R. § 207.20(a). Registration requirements apply to firms engaged in repackaging, relabeling or making any other changes to packaging in furtherance of distribution. *Id.; see id.* § 207.3(a)(8). Registration requirements apply to establishments in foreign countries if drugs are imported or offered for import into the U.S. See *id.* § 207.40(a). Failure to register an establishment or list a drug is a "prohibited act," and the penalties for failure to register are fines, imprisonment or both. *See* Federal Food, Drug, and Cosmetic Act §§ 301(p), 303(a), 21 U.S.C. § 333(a).

9 The reporting requirements are set forth in 21 C.F.R. § 310.305 (for drugs without approved NDA/ANDAs); 21 C.F.R. §§ 314.80, 314.98 and 314.540 (application drug products); 21 C.F.R. § 600.80 (therapeutic biologics) and section 760 of the FDCA (nonprescription products).

10 Investigations Operations Manual, *supra* note 7, Sec 5.1.2.1.

11 *Id.* Sec. 5.5.1.

ence-based decisions and acts promptly to identify, investigate, correct, and prevent manufacturing problems likely to, or [that] have led to, product quality problems."[12]

As described in the FDA IOM, the FDA investigator's authority to inspect is "predicated on specific obligations to the firm . . . It is [the investigator's] responsibility to conduct all inspections at reasonable times and within reasonable limits and in a reasonable manner."[13] Upon arrival at the establishment, the investigator is required to present his or her credentials as well as a Form FDA-482, Notice of Inspection, to the firm's top management official.[14]

The IOM instructs investigators to conduct a preliminary tour of the facility in order to establish the depth of the inspection and the inspection strategy. This visual inspection provides information about firm housekeeping, manufacturing operations and training of staff. During the preliminary tour, investigators are instructed to look for obvious potential product problems.[15] As part of their record review, investigators typically will review complaint files, deviations, and OOS investigations. This review also helps investigators to identify suspect products.[16]

FDA takes the position that it is authorized to take photographs during establishment inspections. There is no explicit authority for photographs under section 704 of the FDCA, and FDA relies on two court cases for its authority, *Dow Chemical v. United States*, 476 U.S. 227 (1986), and *U.S. v. Acri Wholesale Grocery Co.*, 409 F. Supp. 529 (S.D. Iowa 1976).[17] In 2012, the IOM was revised to provide explicit instruction to investigators in the event that a firm refuses to allow photographs:

> [O]btain name and contact information for the firm's legal counsel and advise your district management immediately. If the firm does not have legal counsel on retainer, collect the name and contact information for the most responsible individual. District management will inform their ORA [FDA's Office of Regulatory Affairs] Regional Counselor in the Office of Chief Counsel (OCC) of the situation, and OCC will then contact the firm's legal counsel or most responsible individual to discuss FDA's legal right to take pictures during inspections. OCC will relay the results of this conversation to district management.[18]

As discussed above, a provision of FDASIA enacted in 2012 provides that a drug is adulterated if it has been manufactured, processed, packed, or held in a facility that "delays, denies, or limits an inspection."[19] In FDA's guidance on this subject, FDA takes the position that resisting or limiting photography by an FDA investigator may constitute an unlawful "limitation" of an inspection.[20]

The IOM also provides instructions to investigators regarding affidavits. Section 704 of the FDCA does not provide explicit authority for affidavits, but the IOM informs the investigator that an affidavit can be used

---

[12] *Id.* Sec. 5.5.1.2.

[13] 21 U.S.C. § 374(a)(1)(B).

[14] *Id.*

[15] Investigations Operations Manual, *supra* note 7, Sec 5.1.2.2.

[16] *Id.*

[17] *Id.* Sec. 5.3.4.1.

[18] *Id.*

[19] 21 U.S.C. 351(j).

[20] Guidance for Industry: Circumstances that Constitute Delaying, Denying, Limiting, or Refusing a Drug Inspection (October 2014) at 7.

in court "either to establish federal jurisdiction or fix the responsibility for a violation."[21] The IOM instructs the investigator to have the individual whom FDA proposes as an affiant review the affidavit and make edits, if necessary, before signing. "Mistakes, corrected and initialed by the affiant are an indication [that] he/she has read and understood the statement."[22] The IOM also encourages investigators to obtain handwritten statements that the individual understood the affidavit:

If it appears that the individual will not sign the affidavit, the IOM instructs the Investigator to be persistent, asking the individual to read and edit the affidavit, and to add the following handwritten notation at the bottom of the affidavit: "I have read this statement and it is true, but I am not signing it because ... ."[23]

While the facts of individual situations might differ, legal counsel generally recommends that company representatives not sign or otherwise acknowledge FDA affidavits, which can be found to be admissions against the interests of the company. A manufacturer should consider incorporating requirements related to affidavits into its inspection management procedures, with instruction to personnel not to read, edit, acknowledge, or listen to the contents of an affidavit. The IOM anticipates the possibility that individuals will refuse to sign "upon advice of corporate counsel" or "per corporate policy" and instructs the investigator to document such a reason for refusal on the affidavit.[24]

Inconsistencies found by FDA during an establishment inspection may result in the issuance of a form FDA 483 detailing the FDA investigator's observations of violative conditions. It is important to note that FDA inspections offer a "snapshot" of a firm's CGMP compliance limited to only those operations and information reviewed by the investigators. Thus, the fact that FDA has not issued a Form FDA 483 or has issued a Form FDA 483 with relatively few observations does not necessarily guarantee that the firm is in compliance, particularly for systems that were not audited during the inspection.

Section 5.2.3 of FDA's IOM states that investigators make "every reasonable effort" to discuss observations as they occur or on a daily basis "to minimize surprises, errors and misunderstandings when the 483 is issued."[25] The IOM also informs investigators that firm management may ask questions about observations, request clarification and inform the investigators about corrections that have been made or will be made during the inspection.

During the inspection exit interview, management should take the opportunity to ask additional questions if an observation is not clear. If the firm believes that there are factual mistakes or that an observation is based on a misunderstanding, it is appropriate to respectfully bring this to the attention of the investigator or inspection team.

While there is no formal requirement to respond to a Form FDA 483, establishments that do not respond risk possible administrative or enforcement action. If FDA determines that responses to Form FDA 483 observations and the corrective actions taken by the establishment are not satisfactory, FDA may issue a Warning Letter or take other administrative action, such as placing the company on import alert. Preparing a response

---

21 INVESTIGATIONS OPERATIONS MANUAL, *supra* note 7, Sec. 4.4.8.

22 *Id.* Sec. 4.4.8.1.

23 *Id.*, Sec. 4.4.8.2.

24 *Id.*

25 *Id.* Sec. 5.2.3.

to a Form FDA 483 can be a time-consuming and stressful process. It diverts resources from other activities and, if not managed properly, the process can be overwhelming. However, if a Form FDA 483 response is well-written, factually correct, and complete, it will demonstrate that the firm has taken or is in the process of taking appropriate corrective and preventive action. This can allay FDA's concerns about the firm's compliance, and reflect positively on the firm. If the response is poorly written, contains errors or omits critical information, FDA will not be able to determine that adequate corrective actions have been implemented, which may in turn potentially trigger additional inspections or a Warning Letter.[26] If the firm submits inadequate responses to a Warning Letter, the firm is at risk for seizures, suspension of product approvals, injunctive action or possibly criminal prosecution.

Crafting an acceptable response requires thoughtful analysis and precision. The response is an opportunity to communicate to the agency an understanding of the firm's compliance obligations. It is also an opportunity to demonstrate that the firm has corrected, or will correct, any deficiencies identified in the response and will ensure that the problems do not recur.[27] In drafting the response, the manufacturer should be careful to address FDA's concerns and demonstrate an understanding of the scope of the problem. For example, the manufacturer should determine whether the observation describes an isolated incident or a systemic problem, and whether the problem is limited to a single lot or extends to other lots of the same product or other products. The manufacturer should also determine whether the problem is contained within a particular department or extends to other departments or other facilities.

Second, the firm should conduct a thorough investigation and take appropriate corrective action. Far too often, manufacturers respond to Form FDA 483s by simply defending their actions or relating the history of a problem. The quality unit should thoroughly investigate FDA's concerns and describe the investigation, investigation results, and corrective actions in the response. Responses should be factual, complete, and easy to read. They should include reasonable timelines, address issues item-by-item, and demonstrate an understanding of regulatory requirements. A poorly written or defensive response suggests that the firm does not pay attention to detail and does not understand the significance of the observations.

A common mistake is a firm's promise during the inspection exit interview to submit a response as quickly as possible. FDA requires responses to be submitted within 15 working days, but some manufacturers attempt to respond sooner in order to demonstrate quick action. A better approach is to use the time allowed. A lengthy Form FDA 483 or one involving complicated issues could require more investigation than anticipated. If the manufacturer is unable to complete all corrective actions within 15 working days, FDA expects the manufacturer to explain the reason for the delay and when the corrective actions will be completed. In this circumstance, the manufacturer should commit to provide periodic updates to FDA until its corrective actions are complete.[28]

A firm should set timetables for corrective and preventive actions that demonstrate an appropriate sense of urgency. Commitments should be realistic, and firms must meet their commitments. Failure to take appropriate corrective action to address objectionable conditions will result in repeat observations (often the basis for a Warning Letter or other enforcement action).

---

[26] Warning Letters are discussed in greater detail in Chapter III, *infra*.

[27] Note that in its subsequent inspection, FDA will ensure that the manufacturer has taken the corrective actions described in the FDA Form 483 response and will verify the adequacy of those corrective actions.

[28] Depending on the significance of FDA's observations, manufacturers generally submit updates on a monthly or quarterly basis.

FDASIA granted FDA additional authorities with respect to records inspection and data submissions. Specifically, section 706 of FDASIA modified section 704(a) of the FDCA by allowing FDA to request records "in advance of or in lieu of an inspection."[29] FDASIA also gave FDA authority to require electronic submissions of marketing applications for drugs, biologics, and device products.[30] In addition, FDASIA allows this authority to be exercised through FDA guidance.[31]

FDASIA also included a number of statutory additions to the FDCA that explicitly expanded FDA's permissible interactions with foreign governments. Section 710 of FDASIA granted FDA the ability to protect confidential information obtained from foreign governments against Freedom of Information Act disclosures.[32] In addition, FDA is authorized to enter into agreements with foreign governments and agencies to recognize their establishment inspections if FDA determines the foreign counterparts' inspections are adequate.[33]

## B. Samples

Sample collection is an important part of FDA's regulatory and enforcement activities. Under the inspection provisions of section 702(a) of the FDCA, FDA is granted authority to conduct investigations and collect samples.[34] FDA's IOM contains a comprehensive policy governing sampling protocols.[35] One reason the sampling protocols are so comprehensive is that collected samples may be used as evidence in enforcement actions and FDA wants to ensure samples are collected appropriately and in such a way that they will be admissible in court.[36] As a starting point, the IOM describes "official samples" as those subject to federal jurisdiction, specifically physical samples from a lot that was introduced, delivered, or received in interstate commerce or a lot that was manufactured in a U.S. territory or the District of Columbia.[37] If a lot becomes adulterated or misbranded after its shipment in interstate commerce, i.e., when held for sale, the authority relied upon to sample is section 301(k) of the FDCA.[38] "301(k)" sampling is distinguishable from other official samples in that this type of sampling must have documentation or other evidence of the act that caused the misbranding or adulteration.[39]

Other types of samples include "documentary" samples such as product labels, transit paperwork, or photographs.[40] Often this type of sampling is associated with misbranding charges.[41] "Investigational" samples are collected to support regulatory actions or observations, but are not necessarily collected from a product lot that has entered interstate commerce or is otherwise under federal jurisdiction.[42]

---

29 Food and Drug Administration Safety and Innovation Act (FDASIA), Pub. L. No. 112-144, § 706, 126 Stat. 993, 1067-68 (2012).

30 FDASIA § 1136, 126 Stat. at 1123-24.

31 *Id.*

32 FDASIA § 710, 126 Stat. at 1070-71.

33 FDASIA § 712, 126 Stat. at 1072.

34 Investigations Operations Manual, *supra* note 7, Sec. 4.1.1.1

35 See *id.* Ch. 4.

36 *Id.* Sec. 4.1.2.

37 *Id.* Sec. 4.1.4.

38 *Id.* Sec. 4.1.4.4.

39 *Id.*

40 *Id.*, Sec. 4.1.4.2.

41 *Id.*

42 *Id.* Sec. 4.1.6.

The IOM also advises FDA personnel on their interaction with "dealers" or those in possession of the product lot.[43] For example, dealers may disagree with a proposed sampling protocol on a given lot. In such a case, the IOM advises a compromised, alternative sampling method be employed.[44] In addition, during sampling activities, affidavits may be collected by individuals associated with the sampled products to record material facts.[45]

## C. Field Alert Reports

Once in the stream of commerce, defects in drug products that affect their safety, quality, identify, purity, or potency may at times be identified. These issues, typically identified by those administering the drug product, must be reported to FDA. The source of the defect may stem from product contamination, defects within individual components of the drug product, problems associated with the packaging of the product, problems with the label, etc.[46]

After a drug product has been distributed in interstate commerce, the manufacturer has a continuing reporting obligation. FDA requires manufacturers of approved drug products to report problems quickly, in order to minimize the threat of harm to patients. For this reason, within three working days of the identification of a significant problem with an approved drug, a holder of an approved NDA or ANDA must submit to the district FDA office a Field Alert Report (FAR).[47] The FAR should include information concerning:

- Any incident that causes the drug product or its labeling to be mistaken for, or applied to, another article; and
- Any bacteriological contamination, or any significant chemical, physical or other change or deterioration in the distributed drug product, or any failure of one or more distributed batches of the drug product to meet the specification established for it in the application.[48]

FDA uses FARs as an "early warning system" and monitors the investigation of reported events to ensure that the manufacturer has taken appropriate corrective action.[49] During establishment inspections, investigators will confirm that FARs are being reported properly.

## D. Adverse Event Reports and Consumer Complaints

Holders of NDAs, ANDAs and antibiotic applications, manufacturers of marketed prescription drugs for human use without approved NDAs or ANDAs, compounders operating in registered outsourcing facilities, and manufacturers of approved biologic products are required to report adverse events or adverse experiences

---

43 *Id.* Sec. 4.2.

44 *Id.* Sec. 4.2.2.

45 *Id.* Sec. 4.4.8.

46 See generally Food & Drug Admin., Product Problems, http://www.fda.gov/Safety/MedWatch/HowToReport/ucm053091.htm. Firms can take samples from FDA sampled lots for their own documentation.

47 See Food & Drug Admin., New Drug Application (NDA) and Abbreviated New Drug Application (ANDA) Field Alert Reports, http://www.fda.gov/AboutFDA/CentersOffices/OfficeofMedicalProductsandTobacco/CDER/ucm082083.htm; 21 C.F.R. § 314.81(b) (1).

48 21 C.F.R. § 314.81(b)(1)(i)-(ii).

49 Examples of reportable events include OOS results for products on stability, complaints of contamination and product mix-ups.

(AEs).[50] An AE is any adverse event occurring in a patient after the administration of a drug product, "whether or not considered drug related, including the following: An adverse event occurring in the course of the use of a drug product in professional practice; an adverse event occurring from drug overdose whether accidental or intentional; an adverse event occurring from drug abuse; an adverse event occurring from drug withdrawal; and any failure of expected pharmacological action."[51] AEs can occur in patients or clinical investigation subjects and may be classified as life-threatening, serious or non-serious, expected or unexpected, study-related, possibly study-related, or not-study related.[52] AEs can relate to domestic or foreign reports.

Application holders and manufacturers are typically made aware of AEs via customer complaints. Once received, application holders or manufacturers should identify the following four data elements: i) identifiable patient; ii) identifiable reporter; iii) a suspect drug or biological product; and iv) an AE or fatal outcome.[53] Where these four data elements have been ascertained, applicants and manufacturers must report to FDA AEs in accordance with the classification of the AE. For purposes of classification, FDA gives the initial reporter's terminology deference.[54] Serious and unexpected AEs require the applicant to issue an Alert Report no later than 15 days after the initial receipt of information, conduct an investigation[55] and submit a follow-up report within 15 calendar days of receipt of new information as requested by FDA.[56] AEs are reported to FDA through MedWatch.

## E. Risk-Based Selection of Sites and Inspection Frequency/Rigor

As part of its oversight functions, FDA has sought to prioritize the inspection of domestic pharmaceutical manufacturing sites.[57] The goal of this approach is to achieve a measure of predictability in identifying and conducting inspections that will have the greatest potential of safeguarding the public health.[58] FDA has made clear that "[a] risk-based systems audit approach is recommended in which higher risk, therapeutically significant, medically necessary, and difficult to manufacture drugs are covered in greater detail during an inspection."[59]

Under this risk-based approach, once a firm has developed an established record of compliance, FDA inspections are likely to occur less frequently than for those firms whose compliance history shows continuous

---

[50] See 21 C.F.R. §§ 310.305, 314.80, 314.98, and 600.80.

[51] 21 C.F.R. § 314.80 (defining adverse drug experience and setting forth requirements of reporting adverse drug experiences occurring post-marketing).

[52] *Id.*

[53] See Food & Drug Admin., Guidance For Industry: Postmarketing Adverse Experience Reporting For Human Drug And Licensed Biological Products: Clarification Of What To Report (August 1997), www.fda.gov/downloads/Drugs/ GuidanceComplianceRegulatoryInformation/Guidances/ucm071981.pdf.

[54] *See, e.g.*, 21 C.F.R. § 314.80 (defining a life-threatening adverse drug experience as "[A]ny adverse drug experience that places the patient, in the view of the initial reporter, at immediate risk of death from the adverse drug experience as it occurred, i.e., it does not include an adverse drug experience that, had it occurred in a more severe form, might have caused death."); *see also* Expedited Safety Reporting Requirements for Human Drug and Biological Products, 62 Fed. Reg. 52,237, 52,242 (Oct. 7, 1997) (to be codified at 21 C.F.R. pts. 20, 310, 312, 314, 600) (FDA declines to revise or delete the phrase, "in the view of the initial reporter," when issuing its Final Rule in 1997).

[55] The manufacturer's quality unit should be notified of the adverse event in order to determine whether product quality is implicated.

[56] See 21 C.R.R. § 312.50(1)(i) – (ii).

[57] Food & Drug Admin., Pharmaceutical Cgmps For The 21st Century – A Risk-Based Approach: Final Report at 8 (2004), http://www.fda.gov/downloads/Drugs/DevelopmentApprovalProcess/Manufacturing/QuestionsandAnswersonCurrent GoodManufacturingPracticesCGMPforDrugs/ucm176374.pdf [Fda Final Report].

[58] *Id.*

[59] Investigations Operations Manual, *supra* note 7, Sec. 5.5.1.2.

or repeated violations of CGMPs.[60] A pilot program initiated by the Center for Drug Evaluation and Research (CDER) in 2012 evaluated the quality of drug programs by identifying a number of "risk parameters [a structured dataset] across clinical investigator sites associated with marketing applications."[61]

## F. Regulation of CGMPs for OTC Monograph and Homeopathic Drugs

Over-the-counter (OTC) drug products are subject to nearly all of the CGMP requirements that apply to prescription drug products. There are, however, a few distinctions for OTC products. During the 1978 CGMP rule promulgation, FDA noted that expiration dating for all OTC products "may not be advantageous to consumers" in some situations.[62] Specifically, the CGMP preamble in 1978 stated that FDA would not enforce expiration dating for OTC products without dosage limitations and the products were stable for at least three years (supported by stability data).[63] One of the stated rationales was that this subset of OTC products tended to be consumed relatively quickly.[64] At the time of the 1978 CGMP rule, this exemption was tentative as FDA intended to undergo further consideration of its OTC policy.[65] However, this OTC exemption from CGMP expiration dating is still effective for those OTC products without dosage limitations and sufficient stability.[66] Manufacturers of OTC products that do not require expiration dating must: 1) keep a reserve sample for three years after a batch's distribution,[67] 2) keep records for three years after a batch's distribution[68] and 3) maintain complaint files for three years after a distribution.[69] Lastly, OTC drug products are required to have tamper-resistant packaging.[70] This type of packaging has to be designed in order to inform the consumer of any tampering with an OTC product.[71] The barrier packaging either needs to be made of materials not commonly available or employ an identifying characteristic (e.g., logo) over the barrier of commonly available materials.[72]

Under the FDCA a "drug" includes an article recognized in the official Homeopathic Pharmacopeia of the United States (HPUS).[73] This definition has been in place since the passage of the FDCA in 1938. There are no provisions in the FDCA that create exemptions for homeopathic drugs from the requirements of the Act. However, since homeopathic drugs are non-traditional and not always capable of meeting FDA pre-approval standards, on May 31, 1988, the agency issued Compliance Policy Guidance (CPG) detailing the conditions

---

60 See FDA FINAL REPORT, *supra* note 57, at 8.

61 See FOOD & DRUG ADMIN., GUIDANCE FOR INDUSTRY: PROVIDING SUBMISSIONS IN ELECTRONIC FORMAT – SUMMARY LEVEL CLINICAL SITE DATA FOR CDER'S INSPECTION PLANNING 3 (Dec. 2012), www.fda.gov/downloads/Drugs/ DevelopmentApprovalProcess/FormsSubmissionRequirements/ UCM332468.pdf.

62 Current Good Manufacturing Practice in Manufacture, Processing, Packing, and Holding, 43 Fed. Reg. 45,014, 45,057 (Sept. 29, 1978) (to be codified at 21 C.F.R. pts. 210-11).

63 *Id.*

64 *Id.*

65 *Id.*

66 See 21 C.F.R. § 211.137(h). There is also an enforcement exemption from general CGMP requirements for OTC drug products if all of their components are considered human food. See *id.* § 211.1(c); See also Food & Drug Admin., CPG Sec 480.300 "Lack of Expiration Date of Stability Data." http://www.fda.gov/ICECI/ComplianceManuals/CompliancePolicyGuidanceManual/ucm074410.htm.

67 21 C.F.R. § 211.170(a)(3), (b)(3).

68 *Id.* § 211.180(a).

69 *Id.* § 211.198(b).

70 *Id.* § 211.132.

71 *Id.* § 211.132(b).

72 *Id.*

73 21 U.S.C. § 321(g)(1)(A).

required to legally market a homeopathic drug product.[74] The CPG was last updated by FDA on January 12, 2010. It states that although homeopathic drugs are required to abide by existing regulatory standards for labeling, registration, and certain CGMPs, homeopathic drug products do not have to abide by the requirements for expiration dating (21 C.F.R. § 211.137) or testing and release for distribution (21 C.F.R. § 211.165).[75] In addition there is no requirement to obtain FDA approval prior to marketing homeopathic drug products, or importing them for sale into the United States.[76]

[74] Food & Drug Admin., Sec. 400.400 Conditions Under Which Homeopathic Drugs May be Marketed, in Compliance Policy Guide (revised 1995), http://www.fda.gov/ICECI/ComplianceManuals/CompliancePolicyGuidanceManual/ucm074360.htm.

[75] *Id.*

[76] See *id.*

# Chapter III
# What Are the Consequences of Noncompliance?

The CGMP regulations and FDA's inspection process are designed to ensure that drug manufacturers operate in a state of control. Widespread or persistent CGMP violations indicate a lack of control and may increase the likelihood that a manufacturer will release defective products and harm patients.

The risks associated with systemic or persistent noncompliance are significant. FDA has an array of advisory, administrative and judicial enforcement options: the agency can subject an establishment to repeated inspections, issue Warning Letters, impose import alerts, take legal action to seize product, sue for injunctive relief, or bring a criminal prosecution.

## A. Advisory Actions: Warning Letters, Untitled Letters and Regulatory Meetings

On August 6, 2009, FDA Commissioner Margaret Hamburg launched a new FDA enforcement initiative designed to protect public health. In her speech, the Commissioner outlined six initial steps to make FDA enforcement more effective: 1) establishment of a 15-working-day deadline for Form FDA 483 responses; 2) streamlined process for issuing Warning Letters; 3) closer cooperation with international, state and local regulatory partners on risk mitigation and enforcement strategies; 4) prioritized Warning Letter and enforcement action follow-up; 5) rapid response to significant public health threats; and 6) a formal Warning Letter close-out process.[1] The Commissioner noted that these changes would "help to ensure that violations are taken seriously, that Warning Letters and enforcement actions occur in a timely manner, and that steps are taken to protect consumers in cases where immediate enforcement action is not possible."[2]

FDA has numerous avenues for communicating with regulated industry about regulatory concerns and/or cited violations depending on the severity of the identified concern or violation.

### 1. Warning Letters

FDA issues Warning Letters to notify entities of violations of regulatory significance, (i.e., those that may lead to enforcement action by FDA if not promptly corrected). Warning Letters are designed to notify top management of the severity and scope of the identified violations, achieve voluntary compliance/corrective actions, prevent recurrence, and establish prior notice of violations.[3] Warning Letters are advisory in that they do not commit FDA to

---

[1] Margaret A. Hamburg, Comm'r, Food & Drug Law Inst.: Effective Enforcement and Benefits to Public Health (Aug. 6, 2009) (transcript available at http://www.fda.gov/NewsEvents/speeches/ucm175983.htm).

[2] *Id.*

[3] Food & Drug Admin., Warning Letters, in REGULATORY PROCEDURES MANUAL 4-2, 4-3 (2017), http://www.fda.gov/downloads/ICECI/ComplianceManuals/RegulatoryProceduresManual/UCM074330.pdf [Warning Letters].

take any particular enforcement action.[4] Warning Letters also are not a prerequisite for FDA to take enforcement action.[5] Moreover, FDA will not tolerate continuing violations and does not intend to issue a second Warning Letter to the same establishment for the same or similar violations.[6] If an establishment does not take corrective action, documentation that the agency provided prior notice may strengthen its position in an enforcement action.[7]

For domestic inspections, the recommendation for the issuance of a Warning Letter typically begins at a district office where the letter is drafted within 15 working days after the completion of the inspection, and then circulated to the applicable center for review and concurrence.[8] The center has 15 working days from receipt of the district's Warning Letter recommendation to review and notify the district of its position.[9]

After the Warning Letter is issued, if the district or center finds that the facility's response is inadequate, or if the facility fails to respond to the Warning Letter, the district or the center may initiate follow-up action.[10] If a response to a Warning Letter appears adequate, FDA will verify that commitments have been fulfilled and corrections achieved, typically through a follow-up inspection.[11] Once that verification occurs, if the follow-up does not identify other significant violations, FDA will issue a close-out letter indicating that the Warning Letter issues have been adequately addressed.[12]

Warning Letters have significant consequences. Redacted Warning Letters are posted on FDA's Warning Letter Internet page.[13] Copies of Warning Letters may also be requested from FDA by third parties pursuant to the Freedom of Information Act, 5 U.S.C. § 552.[14] Such publicity may have a negative impact on the manufacturer and public perceptions of its products. Publicity, as described above, may lead to products liability lawsuits from individuals. Indeed, some suits have been based solely on concerns identified in a Warning Letter. In addition, FDA routinely advises other federal agencies of Warning Letters so that the other agencies may take them into account when considering award or renewal of government contracts.

## 2. Untitled Letters

As the name suggests, Untitled Letters are letters issued without the title "WARNING LETTER" in the heading of the letter. Untitled Letters cite "violations that do not meet the threshold of regulatory significance for a Warning Letter."[15] Untitled Letters do not threaten enforcement action, and a follow-up inspection is not

---

4 *Id.* at 4-3. Courts have held that Warning Letters are not final agency action subject to judicial review. See Holistic Candlers and Consumers Ass'n v. Food and Drug Admin., 664 F.3d 940, 942-943 (D.C. Cir. 2012).

5 *Id.*

6 Warning Letters, *supra* note 3, at 4-1 (noting that a second Warning Letter should not be issued to the same firm for the same or similar violation(s)).

7 See Food & Drug Admin., Prior Notice, in Regulatory Procedures Manual 10-7 (2012), http://www.fda.gov/downloads/ICECI/ComplianceManuals/RegulatoryProceduresManual/UCM074292.pdf ("Warning Letters are the principal means by which the agency provides prior notice of violations and of achieving voluntary compliance.").

8 Warning Letters, *supra* note 3, at 4-13.

9 *Id.*

10 *Id.* at 4-14.

11 *Id.*

12 *Id.* at 4-14-4-15.

13 *Id.* at 4-22.

14 *Id.*

15 Food & Drug Admin., Untitled Letters, in Regulatory Procedures Manual 4-33, 4-33 (2012), http://www.fda.gov/downloads/ICECI/ComplianceManuals/RegulatoryProceduresManual/UCM074330.pdf; See also United States ex rel. Colquitt v. Abbott Labs., 864 F. Supp. 2d 499, 515 (N.D. Tex. 2012).

required. FDA also does not routinely advise other federal agencies about Untitled Letters. The recommendation for the issuance of an Untitled Letter begins at the district level and the relevant centers must provide concurrence prior to the issuance of an Untitled Letter. In addition, FDA's procedures do not establish any particular time frame for issuance of an Untitled Letter.

### 3. Regulatory Meetings

Regulatory Meetings are "meeting[s] requested by FDA management at its discretion, to inform responsible individuals or firms about how one or more products, practices, processes, or other activities are considered to be a violation of the law."[16] Regulatory Meetings will not be the first communication of a violation to firms or facilities, but may be used as a follow-up to a Warning Letter, successful implementation of corrective action, or to document further violations that are not deemed to be significant. Regulatory Meetings may also be used to address violations that constitute a health hazard and require immediate attention or in conjunction with the issuance of a Warning Letter.

Firms and responsible officers of firms called to a Regulatory Meeting should be prepared to make a commitment to corrective actions and further and/or increased oversight.

## B. Import Detentions and Import Alerts

FDA has the responsibility to evaluate imported drugs to determine whether they should be admitted into domestic commerce. The FDCA grants FDA authority to refuse admission, in collaboration with United States Customs and Border Protection, of any drug product that is misbranded or adulterated (e.g., in violation of CGMP requirements).[17] If an imported drug product is examined or sampled and believed to be violative, the agency can detain the product.[18] FDA can also issue "Import Alerts" that result in automatic detention without physical examination of specified categories of products when "there exists a history of the importation of violative products."[19]

If FDA detains a product (with or without physical examination), the agency issues a notice to the applicable owner or consignee that describes the alleged violation and provides the opportunity to introduce testimony or other evidence regarding the detained product.[20] If the agency is satisfied that the product is compliant, it is released into domestic commerce, and if not, the importation is refused, unless the owner or consignee can bring the product into compliance.[21] If an Import Alert is involved, the owner or consignee can seek to have the Import Alert lifted. Typically, a minimum of five consecutive non-violative shipments are needed to demonstrate that removal of an Import Alert is justified.

---

[16] See FOOD & DRUG ADMIN., Regulatory Meetings, in REGULATORY PROCEDURES MANUAL 10-7, 10-7 (2012), available at http://www.fda.gov/downloads/ICECI/ ComplianceManuals/RegulatoryProceduresManual/UCM074292.pdf.

[17] 21 U.S.C. § 381(a)(3).

[18] FOOD & DRUG ADMIN., Import Procedures, in REGULATORY PROCEDURES MANUAL 9-1, 9-8 (2017), http://www.fda.gov/downloads/ICECI/ComplianceManuals/RegulatoryProceduresManual/UCM074300.pdf.

[19] *Id.* at 9-8-2.

[20] 21 C.F.R. § 1.94.

[21] 21 U.S.C. §§ 381(a), 381(b).

## C. Recalls

FDA has mandatory recall authority for certain FDCA regulated products (e.g., infant formula, medical devices, food), yet most of its recall efforts are conducted through voluntary requests of manufacturers.[22] While drug product recalls are "voluntary," FDA can usually persuade a manufacturer to recall violative products from the market. If the manufacturer fails to initiate a recall at FDA's request, the agency has the ability to issue a public warning, seize product, or take other enforcement action.[23]

There are "recall classifications" that represent FDA's assessment of the relative hazard presented by a product. A Class I recall occurs when there is a "reasonable probability" that the violative product will result in serious adverse health risk or death.[24] A Class II recall occurs when the violative product "may cause temporary or medically reversible" adverse health impacts or there is a remote risk of serious health consequences.[25] A Class III recall occurs when the violative product is unlikely to cause adverse health outcomes.[26] FDA also has Recall Guidance, which specifies the level or "depth" of recall that may be necessary.[27] The depth of a recall is the level to which notification is required (i.e., wholesale, retail, or consumer). In the case of a product distribution that has reached consumers, FDA advises a "sub-recall" or instructions to retailers to initiate mini-recalls of the distributed product.[28]

## D. Seizures

The FDCA authorizes seizure actions as one type of enforcement action to remedy violations of the statute.[29] A seizure is an in rem action undertaken by the Department of Justice on behalf of FDA for forfeiture of an article "for violation of the law by the article itself."[30] The government can conduct "open-ended" seizures of a specific product or products (all lots) and "lot-specific seizures."[31] A "lot-specific seizure" is specific to one drug product lot or batch. A mass seizure is the seizure of all FDA-regulated products at an establishment. FDA will consider conducting a mass seizure when all of the products were produced under the same conditions (manufacturing processes not in compliance with CGMPs) or in the same environment (insanitary conditions). From FDA's perspective, the mass seizure of inventory from a company's warehouse is a relatively efficient means of preventing adulterated product from being distributed.

Because seizures require the U.S. Marshal to take possession or place products in constructive custody of a U.S. District Court, seizures are not favored when FDA has the ability to exercise control over products through other means. As stated in FDA's Regulatory Procedures Manual:

---

22 Food & Drug Admin., Guidance For Industry: Product Recalls, Including Removals And Corrections (2003), http://www.fda.gov/Safety/Recalls/IndustryGuidance/ucm129259.htm [Product Recalls Guidance].

23 In addition, FDA can formally request a recall under its regulations. 21 C.F.R. § 7.45. Manufacturers are very likely to initiate a recall once FDA has a made a formal request.

24 21 C.F.R. § 7.3(m)(1).

25 *Id.* § 7.3(m)(2).

26 *Id.* § 7.3(m)(3).

27 Product Recalls Guidance, *supra* note 22, at 5.

28 *Id.*

29 21 U.S.C. § 334.

30 Food & Drug Admin., Seizure, in Regulatory Procedures Manual 6-1-1(2012), https://www.fda.gov/iceci/compliancemanuals/regulatoryproceduresmanual/ucm176733.htm [Seizure].

31 *Id.* at 6-1-3.

> Seizure may not be the most appropriate means of control when the Agency has control over products through other means. An example would be halting a sponsor's unlawful shipments of unlicensed biologics due to possible interference with an ongoing attempt to obtain a license.[32]

A company whose products have been seized has three options.[33] First, the company can do nothing, which will result in a default judgment and destruction of the articles. Second, a company can file a claim for the article or articles and enter into a consent decree. As part of the consent decree, the claimant admits that the product is violative and either seeks to bring the article or articles into compliance or have the articles destroyed. The claimant is also required to post a bond, which is generally twice the retail value of the articles. The purpose of the bond is to ensure that the claimant complies with the terms of the consent decree and properly reconditions the seized goods.[34] In recent years FDA has sought to have provisions added to seizure consent decrees that enjoin future unlawful acts by the company and give FDA ongoing supervisory authority over various aspects of the company's operations (like the provisions of consent decrees in injunction cases described below).

A third option is to contest the seizure by denying some or all of government's allegations.[35] In order to contest a seizure, a claimant must file with the court a verified claim demonstrating its interest in the property. At that point the case proceeds like an ordinary civil litigation, with discovery and an eventual trial. If the claimant prevails, the goods are released. If FDA prevails, the goods are either destroyed or reconditioned in a manner acceptable to FDA.

As a practical matter, an open ended seizure or mass seizure of a manufacturer's drug products effectively shuts down the manufacturer's operations because further distribution of product places the company at risk for more seizures. Consequently, a seizure action is not only effective in removing product from the market, but it is also effective in expediting negotiation of a consent decree.

## E. Injunctions

FDA has a number of enforcement tools it can employ to ensure that manufacturers comply with the FDCA. One of the most powerful of these is an injunction.[36] In an injunction action, the Department of Justice (on behalf of FDA) requests a court to order a company to take or refrain from taking specified actions. The purpose of the injunction is to stop current violations of the FDCA and prevent future violations.

Depending on the severity of the action FDA seeks to curtail through the injunction, FDA could seek a temporary restraining order (TRO) or a preliminary injunction.[37] In determining whether a TRO or a preliminary injunction should issue, the court typically will consider four factors:

> (1) whether the plaintiff is likely to succeed on the merits; (2) whether the plaintiff is likely to suffer irreparable injury if the temporary restraining order or preliminary injunction does not issue;

---

32 *Id.* at 6-1-2 (6).

33 *Id.* at 6-1-9.

34 *Id.*

35 *Id.*

36 21 U.S.C. § 332; see also Fed. R. Civ. P. 65.

37 Note that in most cases, a manufacturer will continue to distribute products pending a final ruling from the court. Thus, in situations where the government has a strong interest in quickly removing product from the market, a seizure or import alert could precede an injunction.

(3) whether the balance of equities tips in the plaintiff's favor; and (4) whether the public interest favors granting the preliminary relief requested.[38]

If FDA does not pursue a TRO or preliminary injunction, it seeks a permanent injunction, for which it needs to prove the alleged violations at a trial.

Permanent injunctions are FDA's preferred enforcement action when seeking to secure ongoing control over a company's operations. These permanent injunctions also allow FDA to seek judicial recourse for the enforcement of provisions contained in the order, which may include requirements that a company retain an independent third-party CGMP expert, initiate product recalls and revalidate its manufacturing processes.

While courts have discretion to issue an injunction, it is "an extraordinary and drastic form of relief"[39] and FDA must be able to point to current evidence of continuing violations to prevail. However, if an establishment has a history of violations and has failed to take promised corrective actions, or if a health hazard has been identified, the government is more likely to obtain injunctive relief.

FDA's Regulatory Procedures Manual provides insight about when FDA will typically seek injunctive relief, including when a) health hazards are involved, b) a company is the source of numerous violations, has refused to take corrective actions and other enforcement tools available to FDA are not feasible; and c) a company has a history of violations that have not led to a health hazard, but which the company has failed to correct through voluntary means.[40]

To resolve an enforcement action without the expense and uncertainty of litigation, many manufacturers will voluntarily agree to refrain from certain actions by entering into a consent decree with the government—frequently captioned as a "Consent Decree of Permanent Injunction." Consent decrees are permanent injunctions agreed to by the parties and signed by the court. Consent decrees typically contain extensive audit and reporting requirements and penalties for noncompliance. Consent decrees are also a mechanism through which FDA can seek to impose disgorgement, an equitable remedy that requires a company to pay to the government all gains received while the drug product was in violation of CGMPs, as well as liquidated damages (mentioned above) to be paid for each day a company fails to operate in compliance with CGMP.[41] Consent decrees generally last for a minimum of five years, and establishments are required to petition the court in order to have the decree lifted. If the decree is not lifted it is in place permanently and its obligations typically pass to any successors or assigns of the company that entered into the decree.

In the event that an establishment or company officer fails to comply with the provisions of the decree, there are potential penalties, including FDA's imposition of monetary penalties established in the decree, i.e., liquidated damages, or even penalties for civil or criminal contempt of a court order.

---

[38] See, e.g., Winter v. Natural Resources Defense Council, Inc., 555 U.S. 7, 20 (2008).

[39] Allina Health Servs. v. Sebelius, 756 F. Supp. 2d 61, 65 (D.D.C. 2010).

[40] See Food & Drug Admin., Injunctions, in Regulatory Procedures Manual 6-25, 6-27–6-28 (2015), http://www.fda.gov/downloads/ICECI/ ComplianceManuals/RegulatoryProceduresManual/UCM074317.pdf.

[41] See Eric M. Blumberg, Abbott Laboratories Consent Decree and Individual Responsibility Under the Federal Food, Drug, and Cosmetic Act, 55 Food & Drug L.J. 145 (2000) (detailing the rationale behind incorporating disgorgement and liquidated damages in Consent Decrees for Permanent Injunctions).

Civil contempt sanctions may be imposed by a court to spur compliance with a consent decree. A finding of civil contempt requires clear and convincing evidence that a party violated a court order or consent decree and that the court order was reasonably clear and specific,[42] and not necessarily a determination of whether a party's failure to comply was either willful or intentional, though willful behavior may be the basis for civil contempt.[43]

If the government can establish beyond a reasonable doubt that: 1) there was a violation, 2) of a clear and reasonably specific consent decree requirement or court order, and 3) the violation was willful, a court may hold the party (individual or company) in criminal contempt.[44] To establish willfulness, the government must show that the company acted with deliberate or reckless disregard of the requirement as established by the consent decree or court order or that noncompliance was intentional.[45] In order to avoid being held in criminal contempt, a company or individual must make good faith efforts to comply with the requirements of the decree or order.

## F. Criminal Prosecutions

If a firm's compliance problems are particularly egregious, there is also the risk of criminal liability for the company and individuals employed by the company. A person or company that commits a "prohibited act" under the FDCA (e.g., the introduction into interstate commerce of a drug adulterated by virtue of having been manufactured in violation of CGMP) may be convicted of a crime. Most violations of the FDCA and associated regulations constitute "prohibited acts" if the affected product was shipped or received in interstate commerce.[46]

An entity or individual is subject to felony liability for committing a prohibited act with intent to defraud or mislead, or for committing a prohibited act following a prior FDCA criminal conviction.[47] Penalties for felony violations include imprisonment for up to three years per violation and fines up to $250,000 per violation for an individual and up to $500,000 per violation for an entity.[48]

An entity or individual is subject to misdemeanor liability for committing a prohibited act even without evidence of intentional wrongdoing.[49] Penalties for misdemeanor violations include imprisonment for up to one year per violation and fines up to $250,000 per violation for an individual and up to $500,000 for an entity.[50]

Because criminal liability typically hinges on intentional wrongdoing, the FDCA is unusual in imposing misdemeanor liability without such intent. To protect against unfairness, the statute shields individuals and entities from misdemeanor liability under some circumstances if they act in good faith, including when they

---

[42] See United States v. Century Clinic, Inc., 75 F. Supp. 2d 1127, 1134-35 (D. Nev. 1998).

[43] See *id.* at 1135; See also Lee v. Dep't of Justice, 401 F. Supp. 2d 123, 131 (D.D.C. 2005).

[44] See United States v. NYNEX Corp., 8 F.3d 52, 54 (D.C. Cir. 1993); see also 18 U.S.C. § 401(3).

[45] See United States v. Young, 107 F.3d 903, 909 (D.C. Cir. 1997); In re Holloway, 995 F.2d 1080, 1082 (D.C. Cir. 1993).

[46] See 21 U.S.C. § 331 (listing prohibited acts).

[47] 21 U.S.C. § 333(a)(2).

[48] 21 U.S.C. § 333(a)(2); 18 U.S.C. § 3571.

[49] 21 U.S.C. § 333(a)(1).

[50] 21 U.S.C. § 333(a)(1); 18 U.S.C. § 3571.

receive a guaranty from a supplier affirming that the products at issue are not adulterated or misbranded or in violation of drug approval requirements.[51] FDA's regulations set forth suggested forms of acceptable guaranties.[52]

Misdemeanor liability for individuals under the FDCA is governed by the Supreme Court's decision in *United States v. Park,* 421 U.S. 658 (1975). The so-called *Park* doctrine establishes that an individual can be liable for an FDCA misdemeanor without intentional wrongdoing or even awareness of wrongdoing if the individual has a "responsible relationship" to, or a "responsible share in, a violation of the Act."[53] In *Park,* the Supreme Court upheld the imposition of misdemeanor liability on an executive or manager if he or she "had a responsible relation to the situation and by virtue of his position . . . had . . . authority and responsibility to deal with the situation."[54] If a corporate officer is aware of violations, relying on dependable subordinates to address violations will not be sufficient to insulate the officer from criminal liability.[55] However, a corporate agent may raise as a defense that he or she was "powerless" to prevent and/or correct the alleged violation.[56] Because it is premised upon an individual corporate officer's responsible relationship to the violation, the *Park* doctrine is also referred to as the "responsible corporate officer" doctrine.

In *Park,* Supreme Court held that the government proves its prima facie case of individual criminal liability when:

> it introduces evidence sufficient to warrant a finding by the trier of the facts that the defendant had, by reason of his position in the corporation, responsibility and authority either to prevent in the first instance, or promptly correct, the violation complained of, and that he failed to do so. The failure thus to fulfill the duty imposed by the interaction of the corporate agent's authority and the statute furnishes a sufficient causal link. The considerations which prompted the imposition of this duty, and the scope of the duty, provide the measure of culpability.[57]

In *Park,* the Supreme Court justified this type of individual liability as follows:

> [The FDCA] imposes not only a positive duty to seek out and remedy violations when they occur but also, and primarily, a duty to implement measures that will insure that violations will not occur. The requirements of foresight and vigilance imposed on responsible corporate agents are beyond question demanding, and perhaps onerous, but they are not more stringent than the public has a right to expect of those who voluntarily assume positions of authority in business enterprises whose services and products affect the health and well-being of the public that supports them.[58]

---

[51] 21 U.S.C. § 333(c)(1), (c)(2).

[52] 21 C.F.R. § 7.13.

[53] *Park,* 421 U.S. 658, at 673-74.

[54] *Id.* at 674 (internal quotations omitted) (finding that the president of a large national food store chain was criminally liable for the insanitary conditions identified by FDA at two of the company's warehouses).

[55] *Id.* at 677 (because the president was on notice of insanitary conditions at one of the company's warehouses, his subsequent reliance on a corporate structure where dependable subordinates would also handle the remediation of the same violations identified by FDA in a separate warehouse was sufficient to demonstrate "that he could not rely on his system of delegation to subordinates to prevent or correct insanitary conditions at [the company's] warehouse, and that he must have [therefore] been aware of the deficiencies of this system before the [subsequent] violations were discovered"). Notably, the court also acknowledged that while Congress may have imposed a duty on corporate agents that requires "the highest standard of foresight and vigilance" it does not require that which is objectively impossible.

[56] *Id.* at 673. To date, this defense has not been argued successfully in any published reports.

[57] *Id.* at 673-74.

[58] *Id.* at 672 (emphasis added).

In July 2012, FDA amended its Regulatory Procedures Manual to include Section 6-5-3, "Special Procedures and Consideration for Park Doctrine Prosecutions." The Regulatory Procedures Manual states that under *Park*:

> [A] responsible corporate official can be held liable for a first time misdemeanor (and possible subsequent felony) under the Federal Food, Drug, and Cosmetic Act ("the Act") without proof that the corporate official acted with intent or even negligence, and even if such corporate official did not have any actual knowledge of, or participation in, the specific offense. A Park Doctrine prosecution, for the purposes of this section, refers to a recommended prosecution of a responsible corporate official for a misdemeanor violation of the Act.
>
> Misdemeanor prosecution under the Act can be a valuable enforcement tool. Such prosecutions are referred to the Department of Justice. Once a person has been convicted of a misdemeanor under the Act, any subsequent violation of the Act is a felony, even without proof that the defendant acted with the intent to defraud or mislead. Misdemeanor prosecutions, particularly those against responsible corporate officials, can have a strong deterrent effect on the defendants and other regulated entities. In some cases, a misdemeanor conviction of an individual may serve as the basis for debarment by FDA.[59]

In light of the fact that failure to comply with CGMP requirements renders a drug product adulterated, and that reliance on a corporate structure with "dependable subordinates" or corporate agents will not, standing alone, insulate a corporate official from criminal liability, corporate officers should ensure that all complaints of adulteration or violations of the FDCA are properly identified, addressed, corrected and monitored to prevent recurrence. Indeed, as the CGMP regulations make clear, firms must have in place procedures to ensure that responsible corporate officers are aware of complaints, returned drug products, drug products that have been improperly stored, as well as other CGMP investigations.[60]

A recent case underscored the severity of the consequences facing executives who admit responsibility for, and acknowledge they had the authority to prevent illegal activity, but failed to do so: namely, exclusion from further participation in federal healthcare programs. In *Friedman v. Sebelius*, the D.C. Circuit held that the Department of Health and Human Services properly excluded three executives of Purdue Frederick Company after they pleaded guilty to "responsible corporate officer" misdemeanors.[61] Although exclusion was limited to misdemeanor convictions "relating to fraud," and the officers' misdemeanor pleas did not admit to fraudulent conduct, the court held that the fraudulent conduct of others at Purdue—which the responsible corporate

[59] Food & Drug Admin., Prosecution, in Regulatory Procedures Manual 6-48 (2012), http://www.fda.gov/downloads/ICECI/ComplianceManuals/RegulatoryProceduresManual/UCM074317.pdf. In the past few years, there has been an increase in the use of the "responsible corporate officer" doctrine articulated in *Park* to prosecute executives in the pharmaceutical and medical device industries. In 2007, three senior executives from the Purdue Frederick Company—the former CEO, the executive vice president/chief medical officer and the executive vice president/chief legal counsel—all pleaded guilty to a misdemeanor count based on their status as responsible corporate officers based on the company's sales and marketing efforts to downplay the addictive characteristics of OxyContin. See United States v. Purdue Frederick Co., Inc., 495 F. Supp. 2d 569, 570 (W.D. Va. 2007). None of these officers was charged with a knowing violation. In 2009, the government charged medical device manufacturer Synthes for, among other things, shipping adulterated and misbranded products, and for false statements to government officials. Four former executives were charged with a single misdemeanor count of introducing adulterated and misbranded products in interstate commerce under the responsible corporate officer doctrine from *Park*. See Indictment, United States v. Norian Corp., No 09-0403 (E.D. Pa. June 16, 2009) (all four pleaded guilty to a misdemeanor count of introducing a misbranded product).

[60] See 21 C.F.R. § 211.180(f) ("Procedures shall be established to assure that the responsible officials of the firm, if they are not personally involved in or immediately aware of such actions, are notified in writing of any investigations conducted under 211.198, 211.204, or 211.208 of these regulations, any recalls, reports of inspectional observations issued by the Food and Drug Administration, or any regulatory actions relating to good manufacturing practices brought by the Food and Drug Administration.").

[61] Friedman v. Sebelius, 686 F.3d 813 (D.C. Cir. 2012).

officers failed to prevent—meant that the officers' misdemeanor pleas related to fraud. The exclusions of the officers effectively ended their careers in the healthcare industry.

## G. False Claims Act and Qui Tam Investigations

CGMP violations may also form the basis for qui tam actions and investigations by the Department of Justice and/or the Department of Health and Human Services (DHHS) Office of the Inspector General even where a firm may already have entered into a consent decree with FDA. Qui tam lawsuits are civil actions brought by a private citizen, denominated a "relator" (often an employee with inside information of fraud) on behalf of the United States under the False Claims Act (FCA).[62] The FCA provides, among other things, a claim for civil damages against a person who knowingly presents or causes to be presented a false or fraudulent claim for payment or approval to a federal program. Civil penalties under the FCA are three times the amount of damages that the federal government sustains plus $5,000 to $10,000 per claim (adjusted periodically for inflation). A relator in a successful qui tam action will receive a percentage (up to 30 percent) of any recovered funds. The relator who brings the suit files the qui tam complaint under seal in order to give the government an opportunity to investigate the allegations and determine whether it will intervene in the action or permit the relator to move forward with the action on his or her own. Once the government has made a determination regarding whether to intervene, the suit is unsealed and notice is given to the defendant. To date at least one CGMP qui tam investigation has resulted in a recovery of $750 million by the government. In that case, the relator received approximately $96 million.[63] Relators and the Department of Justice have been increasingly creative with respect to how violations of regulatory requirements, including CGMP violations, by firms can cause false claims to be submitted to the federal government. Firms can expect that this trend will continue and that there will be an increasing number of FCA qui tam actions brought based on CGMP violations.

## H. Other Consequences

In addition to enforcement actions and other direct consequences of a CGMP or other FDCA violation, there are indirect consequences that can have a significant impact on a company. An FDA enforcement action can lead to or cause:

- adverse publicity—resulting in a loss of consumer confidence and decreased stock value;
- delayed product approvals—when facilities are found noncompliant, pending applications and manufacturing supplements may not be approved;
- business uncertainty, which can occur for an extended time as violative conditions are remedied;
- legal and consultant fees to remediate and address advisory or enforcement actions; and
- increased product liability litigation, consumer class actions and shareholder derivative suits.[64]

---

[62] 31 U.S.C. §§ 3729-3733.

[63] See 19 FDA Enforcement Manual Newsl. 9, GSK Agrees to Pay $750 Million to Resolve Liability for Manufacturing Deficiencies, Substandard Drugs (Dec. 2010).

[64] Eric M. Blumberg, Compliance with the FDCA B Why Bother?, PHARMA, BIOTECH & DEVICE COLLOQUIUM (June 5, 2005), http://www.ehcca.com/presentations/pharmacolloquium2/blumberg_h1.pdf.

# I. Certificates of Export/Government Contracts

## 1. Export Certificates

An FDA export certificate is a document that the agency prepares to communicate information regarding the regulatory status of a drug or other FDA-regulated product.[65] FDA is required to issue export certificates under section 801(e)(4) of the FDCA.[66] The certificates can attest that a product meets FDA requirements (such as CGMPs) or can attest that the product can be legally exported under the FDCA.[67] There are several categories of export certificates, such as Certificate of a Pharmaceutical Product (used to allow an importing country to review a drug approval), Certificate to Foreign Government (which states that a drug product can be legally marketed in the United States) and Certificate of Exportability (used for drug products not meeting United States requirements but allowed for export).[68] Export certificates cannot be obtained for products manufactured in a facility noncompliant with CGMPs related to the product unless under a Certificate of Exportability.

With respect to the Certificate of Exportability there are two principal statutory provisions of the FDCA that would allow an otherwise violative drug to be exported. Under section 801(e) of the FDCA an approved drug product will not be considered misbranded/adulterated if the product: 1) meets the foreign purchaser's specifications, 2) complies with the laws of the importing country, 3) has packaging signifying that it is intended for export and 4) is not sold in domestic commerce.[69]

For unapproved products and approved products being exported for unapproved uses, section 802 of the FDCA establishes an export pathway. All unapproved products must satisfy general requirements including the requirement to be in "substantial conformity" with CGMPs.[70] FDA has interpreted that term not to mean full compliance, however an exporting firm should have passed its most recent CGMP inspection or rectified previous violations.[71] In addition, the product must comply with section 801(e)(1), i.e., it must meet the specifications of the foreign customer, not be in conflict with laws of the foreign country to which the product is being exported, be labeled as intended for export and not be distributed in U.S. interstate commerce.[72] It should be noted that under FDA's Export Guidance, a product not CGMP compliant can only be exported if the purchaser "specifically requested" the violative modification to the product's manufacturing.[73]

Once the general requirements are satisfied there are three pathways to product export. First, the drug product can be exported if it has a valid market authorization from a listed country.[74] In this case FDA approval is not required for export. A second approach, covering non-listed countries, involves: 1) a firm showing it has

[65] Food & Drug Admin., Guidance For Industry: FDA Export Certificates (2005), http://www.fda.gov/ RegulatoryInformation/Guidances/ucm125789.htm [Exports Guidance].

[66] 21 U.S.C. § 381(e)(4).

[67] *Id.* § 381(e)(4)(A)(ii).

[68] Exports Guidance, *supra* note 65.

[69] 21 U.S.C. § 381(e)(1).

[70] 21 U.S.C. § 382.

[71] Exports Guidance, *supra* note 65.

[72] 21 U.S.C. § 381(e)(1). In addition its labeling must comply with the listed country's rules (if applicable). *Id.*

[73] Exports Guidance, *supra* note 65.

[74] Listed countries are Australia, Canada, Israel, Japan, New Zealand, Switzerland, South Africa or any member nation in the European Union or the European Economic Area. 21 U.S.C. § 382.

a valid marketing authorization from the importing country, 2) FDA determination that the importing country has standards related to safety and efficacy review, CGMPs, reporting of adverse events, and requires labeling and promotion in line with a product's approval.[75] Another alternative is for an exporter to: 1) certify the drug cannot meet the other approval pathways, 2) provide credible scientific evidence to show the drug is safe and effective for indicated use, 3) demonstrate that the importing country has requested the drug's exportation and is aware of its unapproved status, and 4) have the importing country concur with exporting firm's evidence of the drug being safe and effective for indicated use.[76]

### 2. Government Contracts

With respect to government contracts, FDA has assumed a quality assurance role in reviewing CGMP compliance.[77] The agency participates in the Government-Wide Quality Assurance Programs for drug product contracts initiated by the Department of Defense and Veterans Administration.[78] FDA's satisfactory CGMP assessment of manufacturers is a condition of these drug product contracts.[79] A Warning Letter or enforcement action can prompt a contracting agency to deny a government contract, terminate, or decide against renewal of an existing contract.

## J. Application Disapproval/Revocation

Section 505(e) of the FDCA gives FDA discretion to withdraw a drug approval, after providing the applicant with due notice and an opportunity for a hearing, if FDA determines that there are CGMP violations that the applicant has not remedied following notice and an opportunity to cure.[80] However, these procedures are rarely invoked by FDA on CGMP grounds.

Under 21 C.F.R. § 314.200(a), FDA is required to provide a notice of opportunity for hearing if it decides to withdraw the approval of an application. FDA must publish the notice in the Federal Register and give affected parties 30 days after the date of publication to file a written notice of participation and a request for a hearing. FDA provides the notice to applicants by registered or certified mail.[81]

Within 60 days of publication (unless a different time period is specified in the Notice of Opportunity for Hearing (NOOH)), the applicant must submit studies used to justify a hearing.[82] The applicant is allowed to incorporate by reference raw data underlying a study, if the applicant submitted the data previously as part of its application.[83]

---

75 *Id.*

76 *Id.*

77 Food & Drug Admin., Sec. 400.200 Consistent Application of CGMP Determinations, in Compliance Policy Guide (1995), http://www.fda.gov/ICECI/ComplianceManuals/CompliancePolicyGuidanceManual/ucm074356.htm.

78 *Id.*

79 *Id.*

80 See 21 U.S.C. § 355(e); see also John D. Copanos and Sons, Inc. v. Food and Drug Admin., 854 F.2d 510 (D.C. Cir. 1988) (noncompliance with CGMPs is a basis for the withdrawal of a manufacturer's application).

81 21 C.F.R. § 314.200(b).

82 2 *Id.* § 314.200(c)(1).

83 *Id.*

FDA will not consider data or analyses submitted after 60 days unless they are from well-controlled studies begun before the NOOH publication date and the study results were not available at that time.[84] FDA will consider other studies on the basis of inadvertent omission and hardship.[85] The applicant is also required to provide a list of all studies in progress, if the applicant intends to later rely on the study results to support the request for a hearing.[86]

The applicant must submit an analysis of the data supporting the hearing request, require a financial certification or disclosure statement as required by 21 C.F.R. Part 54, and provide FDA's preferred format for submission of clinical data.[87]

Upon receipt of a request for hearing, CDER is required to prepare 1) an analysis of the request for hearing and 2) a proposed order approving or denying the request.[88] The request, along with the analysis and proposed order, must be submitted to the Office of the Commissioner for review and decision.[89] If CDER has recommended denying the request, no CDER representatives are allowed to participate in the review and decision. If CDER recommends that a hearing be granted, CDER representatives may participate or advise in the review and decision.[90]

Under 21 C.F.R. 314.200(g), FDA will grant a hearing if the applicant demonstrates that there is a "genuine and substantial issue of fact." Allegations or denials are not sufficient to prevail on a request for a hearing. Moreover, "if it conclusively appears from the face of the data, information and factual analyses . . . that there is no genuine and substantial issue of fact that precludes . . . withdrawal of approval," the Commissioner must enter summary judgment against the applicant.[91] The Commissioner's order of summary judgment must set forth detailed findings and conclusions, and must specify why each study submitted by the applicant fails to meet the requirements of the FDCA and FDA regulations, or does not raise a "genuine and substantial issue of fact."[92]

If CDER believes that summary judgment is warranted, it will serve, by registered mail, a proposed order denying the hearing.[93] The person requesting the hearing has 60 days from receipt of the proposed order to respond with "sufficient data, information and analyses" to demonstrate that there is a genuine and substantial issue of fact, and that a hearing is justified.[94]

If a review of the data shows that the grounds for the NOOH are not valid (e.g., the applicant provides substantial evidence that the product is safe and effective), the Commissioner will issue summary judgment for the applicant and rescind the NOOH.[95] If the Commissioner grants a hearing, the hearing will begin within 90 days of the expiration of the time for requesting the hearing (generally 60 days after the publication of the

---

84 *Id.* § 314.200(c)(2).
85 *Id.*
86 *Id.*
87 21 C.F.R. § 314.200(d)(3).
88 *Id.* § 314.200(f).
89 *Id.*
90 *Id.*
91 *Id.* § 314.200(g)(1).
92 *Id.*
93 *Id.* at § 314.200(g)(2).
94 *Id.* § 314.200(g)(3).
95 *Id.* § 314.200(g)(4).

NOOH) and as soon as practicable in the case of the withdrawal of approval.[96] The Commissioner is required to grant a hearing if there is a genuine and substantial issue of fact or if the Commissioner believes that a hearing would be in the public interest.[97]

Under 21 C.F.R. 314.200(h), if the opportunity for a hearing is waived or a hearing is denied, FDA will issue a notice of withdrawal and declare that the products are unlawful. The Commissioner is allowed to defer or stay the action pending a ruling on another request for hearing, a related hearing, or other administrative or judicial proceeding.[98]

If a hearing does occur, FDA conducts section 505(e) withdrawal proceedings under its procedural regulations for a "formal evidentiary public hearing."[99] Among other things, these regulations provide for discovery of certain FDA records and a trial-type evidentiary hearing before a neutral agency Presiding Officer involving testimony by, and cross-examination of, witnesses.[100]

## K. Application Integrity Policy

The Application Integrity Policy (AIP) is FDA's internal policy aimed at verifying the integrity of the information and data provided by applicants through applications submitted to FDA for review and approval.[101] Specifically, the AIP is concerned with the submission of fraudulent data: "The policy sets forth FDA's general approach regarding applicants that seek to subvert the agency's review and approval processes for premarket applications. The policy also outlines recommended corrective actions by which applicants may seek to restore FDA's confidence in their integrity and permit the agency to proceed with substantive scientific review of their premarket applications."[102] Data may be deemed fraudulent where it is unreliable whether due to negligence, incompetence, or other like reason.[103] If FDA deems it necessary to invoke the AIP, it may delay or prevent the approval of a submission. Accordingly, it is incumbent on firms submitting data to ensure the integrity of such prior to submission.

---

96 *Id.* § 314.200(g)(5).

97 *Id.* § 314.200(g)(6).

98 21 CFR § 314.000.

99 21 C.F.R. §§ 10.50(c)(16), 12.1(a).

100 See generally 21 C.F.R. part 12.

101 See Food & Drug Admin., Inspections, Compliance, Enforcement, And Criminal Investigations, http://www. fda.gov/ICECI/EnforcementActions/ApplicationIntegrityPolicy/ucm072628.htm [AIP].

102 See Fraud, Untrue Statements of Material Facts, Bribery, and Illegal Gratuities; Final Policy, 56 Fed. Reg. 46,191 (Sept. 10, 1991).

103 See AIP, *supra* note 104, at 10. ("A wrongful act includes, but is not limited to, submitting a fraudulent application, offering or promising a bribe or illegal gratuity, or making an untrue statement of material fact. A wrongful act also includes submitting data that are otherwise unreliable due to, for example, a pattern of errors whether caused by incompetence, negligence, or a practice such as inadequate standard operating procedures or a system-wide failure to ensure the integrity of data submissions. Usually, when the AIP is invoked, there are two or more applications affected.").

# Chapter IV
# Recent CGMP Enforcement Activity

## A. The Pattern of Enforcement

While FDA's inspectional and compliance resources are heavily focused on drug CGMP issues, the vast majority of drug CGMP disputes with FDA are resolved without recourse to formal enforcement actions such as seizures, injunctions, or criminal charges. For example, in fiscal year 2016, FDA's Center for Drugs initiated only one seizure action and three injunctive actions addressing regulatory violations of any kind (not just CGMPs).[1] Nonetheless, when FDA does decide to pursue CGMP violations in a formal enforcement action, the consequences often are very significant. Once initiated, the enforcement action can sweep in statutory and regulatory violations unrelated to CGMPs (or even unrelated to the FDCA), both civil and criminal liability, substantial financial payments, and onerous consent decree obligations. We describe below three significant CGMP enforcement actions that the Department of Justice has cited to illustrate the breadth and depth of possible enforcement activity.[2]

## B. GlaxoSmithKline, PLC and its Subsidiary SB Pharmco Puerto Rico Inc.

In 2010, GlaxoSmithKline PLC and its subsidiary, SB Pharmco Puerto Rico Inc., settled a major drug CGMP enforcement action relating to SB Pharmco's Cidra, Puerto Rico manufacturing site. The violations claimed by FDA and the Department of Justice included failure to ensure that finished products were free from contamination by microorganisms, distribution of products with the potential to lack therapeutic effect, distribution of products with the potential to lack a controlled release mechanism, and distribution of products that did not always contain the FDA-approved mix of active ingredients. The Department of Justice concluded that the departure of key employees with responsibility for drug safety contributed to the conduct that led to the enforcement action.[3]

SB Pharmco pleaded guilty to a felony violation of the FDCA and agreed to pay a criminal fine of $150 million, which included a $10 million asset forfeiture. GlaxoSmithKline entered into a civil settlement agreement in which it agreed to pay an additional $600 million to resolve claims that it caused false claims to be submitted to the government in connection with the deficient drugs described above.[4]

---

1 2016 FDA Enforcement Statistics, https://www.fda.gov/downloads/ICECI/EnforcementActions/UCM540606.pdf.

2 Speech of Deputy Assistant Attorney General Maame Ewusi-Mensah Frimpong at the 2013 CBI Pharmaceutical Compliance Congress, https://www.justice.gov/opa/speech/deputy-assistant-attorney-general-maame-ewusi-mensah-frimpong-speaks-2013-cbi.

3 *Id.*

4 Press Release, GlaxoSmithKline to Plead Guilty & Pay $750 Million to Resolve Criminal and Civil Liability Regarding Manufacturing Deficiencies at Puerto Rico Plant, https://www.justice.gov/opa/pr/glaxosmithkline-plead-guilty-pay-750-million-resolve-criminal-and-civil-liability-regarding.

## C. Genzyme Corporation

In 2010, Genzyme Corporation entered into a consent decree relating to CGMP violations at a manufacturing facility in Alston, Massachusetts. FDA's concerns included drugs contaminated with metal, fiber, rubber, and glass particles, and a viral contamination in a bioreactor.[5] FDA alleged 49 separate deviations from CGMP requirements, including failure to establish an adequate quality control unit; failure to establish appropriate written procedures designed to prevent microbial contamination; failure to investigate out of specification product; contamination of drugs with foreign matter; and low fill volumes for drugs that affected their strength and potency.[6]

In the consent decree, the company agreed to pay $175 million to the government in "equitable disgorgement." The company also agreed to extensive revisions in its manufacturing processes and procedures under specified time frames enforceable through additional monetary penalties.[7]

## D. Ranbaxy Laboratories Limited and its Subsidiary Ranbaxy USA Inc.

In 2012 and 2013, Ranbaxy USA Inc., settled a major CGMP enforcement action relating to conditions at manufacturing facilities in India owned and operated by its parent company Ranbaxy Laboratories Limited. The enforcement action followed almost a decade of CGMP deficiencies documented by FDA, three FDA warning letters and an import alert barring importation of more than 30 drugs into the United States. The CGMP issues uncovered by FDA included incomplete testing records; failure to investigate evidence indicating that drugs did not meet their specifications; failure to prevent cross-contamination between penicillin drugs and non-penicillin drugs; failure to have adequate procedures to prevent contamination of sterile drugs; and inadequate stability testing.[8]

To settle the matter, Ranbaxy first entered into a consent decree that the Department of Justice described as "unprecedented in scope," among other things because it was "groundbreaking in its international reach" in requiring fundamental changes to manufacturing plants in both the United States and India.[9] Ranbaxy later reached criminal and civil settlements including what the Department of Justice described as "the nation's largest financial penalty paid by a generic pharmaceutical company for FDCA violations."[10] To resolve criminal

---

5 Press release, Genzyme Corp. Signs Consent Decree to /Correct Violations at Allston, Mass., Manufacturing Plant and Give up $175 Million in Profits, available at https://www.fda.gov/NewsEvents/Newsroom/PressAnnouncements/ucm213212.htm.

6 Complaint, United States v. Genzyme Corporation et al., Case No. 1:10-cv-10865 (D. Mass. 2010).

7 Press release, Genzyme Corp. Signs Consent Decree to /Correct Violations at Allston, Mass., Manufacturing Plant and Give up $175 Million in Profits, available at https://www.fda.gov/NewsEvents/Newsroom/PressAnnouncements/ucm213212.htm.

8 Press Release, U.S. Files Consent Decree for Permanent Injunction Against Pharmaceutical Ranbaxy Laboratories, available at https://www.justice.gov/opa/pr/us-files-consent-decree-permanent-injunction-against-pharmaceutical-ranbaxy-laboratories.

9 *Id.*

10 Press Release, Generic Drug Manufacturer Ranbaxy pleads Guilty and Agrees to Pay $500 Million to Resolve False Claims Allegations, CGMP Violations and False Statements to the FDA, available at https://www.justice.gov/opa/pr/generic-drug-manufacturer-ranbaxy-pleads-guilty-and-agrees-pay-500-million-resolve-false.

liability, the company paid $150 million and pleaded guilty to seven felony counts charging introduction of adulterated drugs, failure to file timely field alert reports, and false statements in annual reports to FDA. To resolve civil liability, the company paid $350 million in connection with claims that Ranbaxy had caused health care programs to submit false claims to Medicare and Medicaid.[11]

[11] *Id.*

# Chapter V
# How Do We Remain "Current" with CGMPs?

As discussed earlier, FDA assesses CGMP compliance primarily through establishment inspections. Those inspections represent "snapshots" in time, and should not be relied upon by the establishment as the only measure of a firm's compliance.

In order to ensure that it remains current with CGMPs, a firm should establish and maintain an effective compliance program. At an operational level, this means that quality must be built into products and if problems arise they must be investigated thoroughly and corrected promptly and effectively so they do not recur. In order to have an effective compliance program, the firm's senior management must develop and maintain high standards of quality and compliance at every level of the organization.

To meet these standards, senior management must ensure that properly qualified and trained employees comply with all applicable FDA requirements. Through strong leadership, well-articulated responsibilities and an emphasis on personal accountability, an organization will be able to achieve and sustain compliance with CGMPs.

A comprehensive written quality and compliance plan is a key component of a formal compliance program. Such a program ensures that the quality and compliance plan is promoted, enforced and monitored, and that senior management exercises appropriate oversight and control of all facilities. Another important aspect of the plan should be a strong audit program, with regularly scheduled internal audits and periodic audits by independent experts.[1] Outside audits should include the review of out-of-specification (OOS) investigations,

---

[1] Pursuant to CPG Sec. 130.300 "FDA Access to Results of Quality Assurance Program Audits and Inspections" FDA will not review or copy audit reports, but will verify that such audits have been performed and documented, and that corrective actions have been taken to address audit findings. Food & Drug Admin., Sec. 130.300 FDA Access to Results of Quality Assurance Program Audits and Inspections, in Compliance Policy Guide (2007), http://www.fda.gov/ICECI/ComplianceManuals/CompliancePolicyGuidanceManual/ucm073841.htm. Note that on occasion, external audits are conducted by experts working at the direction of outside counsel in anticipation of litigation. An investigation conducted by outside counsel that would not be conducted in the ordinary course of business, for the purpose of seeking legal advice, is privileged. Ziner v. Cedar Crest College, No. 04-3491, 2006 U.S. Dist. LEXIS 34858, at *8 (E.D. Pa. May 30, 2006); Diversified Indus., Inc. v. Meredith, 572 F.2d 596, 610 (8th Cir. 1978) (special investigation by outside counsel into corporate bribes was privileged when conducted for the sake of legal advice). The work product doctrine only protects investigations conducted for the primary motive of anticipated litigation. Elec. Data Sys. Corp. v. Steingraber, No. 4:02-CV-223, 2003 U.S. Dist. LEXIS 11818 (E.D. Tex. July 9, 2003) (no work product protection accorded where the primary purpose was determining employment termination); Caremark, Inc. v. Affiliated Comp. Servs., Inc., 195 F.R.D. 610, 614 (N.D. Ill. 2000); Long v. Anderson Univ., 204 F.R.D. 129, 136 (S.D. Ind. 2001) (no work product doctrine protection accorded to insurance carrier investigation). On the other hand, draft submissions to FDA and memoranda of interviews from the related corporate investigation have been held to be privileged work product. In re Grand Jury Subpoena Dated Nov. 8, 1979, 622 F.2d 933, 935 (6th Cir. 1980). However, the fact that an investigation was conducted is not protected by the work product doctrine. Hickman v. Taylor, 329 U.S. 495, 512-13 (1947). Further, the underlying facts that existed prior to the investigation are not made privileged by the fact that those facts are revealed to an attorney. Alexander v. FBI, 186 F.R.D. 102, 111 (D.D.C. 1998).

nonconforming product reports, complaint investigations, CAPAs, Form FDA-483 observations, Warning Letters and prior audits.[2]

The plan should also provide the following: 1) a written roadmap to achieve sustained compliance and address risks associated with FDA-regulated activities, 2) education of employees whose jobs have or could have an impact on the organization's compliance profile, 3) a procedure to ensure that the organization's auditing and reporting functions will measure the effectiveness of the plan, 4) a mechanism to remedy problems as quickly and efficiently as possible with thorough investigations and effective solutions, and 5) enforcement and disci-plinary consequences designed to ensure that employees take seriously their compliance responsibilities.

In addition, the plan should emphasize strong supplier management. As pharmaceutical manufacturing supply chains become longer and more globalized, FDA investigators will scrutinize supply chain management even more carefully than before. It is also likely that FDA will look to its well-established standards for supply chain management for medical devices, which are set forth in FDA's quality system regulation for purchasing controls,[3] to enforce requirements for drug manufacturers.

Firms should incorporate explicit supply chain management requirements in their quality and compliance plans. At a minimum, firms should ensure that suppliers are familiar with FDA regulations and other quality systems requirements, and that suppliers understand that if they are unwilling to conform to the finished prod-uct manufacturer's expectations, they will be disqualified.

The manufacturer should have supply agreements and quality agreements that are tailored to the transac-tion, the work to be performed and the nature of the components or products being supplied. Moreover, the supply agreement should refer specifically to the quality agreement and incorporate its terms by reference. The quality agreement should establish specifications and requirements consistent with CGMP regulations, require risk-based supplier audits, give the finished product manufacturer the option to provide on-site oversight as necessary and clearly articulate change control procedures to ensure that the supplier provides prompt notifi-cation of all changes or problems that may have an impact on the product (e.g., what might seem to be minor adjustments that may, in fact, have an adverse impact on the finished product). Note that all suppliers should be evaluated in terms of risk, and greater oversight should be required for a sole source supplier or a critical component supplier.

Many firms have separate compliance programs that relate to other regulated areas, such as advertising and promotion of drug products and federal healthcare program requirements related to fraud and abuse. Typically, these compliance programs are highly visible to staff and to the public. We recommend linking

---

[2] In a recent speech, Deputy Assistant Attorney General Maame Ewusi-Mensah Frimpong encouraged companies to conduct independent audits or otherwise have outside experts evaluate their compliance with CGMPs in order to "make sure that your systems are not masking any problems that exist." Maame Ewusi-Mensah Frimpong, Deputy Assistant Attorney Gen., Dep't of Justice, Speech at the 2013 CBI Pharmaceutical Compliance Congress (Jan. 29, 2013).

[3] See 21 C.F.R § 820.50. FDA's purchasing control regulation requires medical device manufacturers to 1) establish specific requirements, including quality requirements for all suppliers, contractors, and consultants; 2) evaluate potential suppliers based on their qualifications and ability to meet the manufacturer's requirements and document the evaluation; 3) define the level and nature of control over supplier and product; 4) maintain an acceptable supplier list; and 5) maintain detailed purchasing documents that clearly define specified requirements, including requirements related to quality, for products and services. *Id.* The regulation also states that, where possible, suppliers are to notify the manufacturer of changes in the product or services so the manufacturer can determine whether the changes affect the quality of the finished device. *Id.* at § 820.50(b).

the CGMP quality and compliance program to other firm compliance programs so that the program receives the same level of support from senior management, communications to staff are consistent and the CGMP program is equally visible to staff and the public.[4]

In conclusion, the most effective way to remain "current" with CGMPs is to have strong senior management support, evidenced by a formal compliance program that establishes the foundation for a culture of compliance, quality and continuous improvement.

4 The GlaxoSmithKline Corporate Integrity Agreement was the first to link CGMP problems to the submission of false claims for adulterated drugs. Wendy Weiss, an assistant U.S. attorney in the U.S. Attorney's Office in Los Angeles and the chief of the Office's Civil Fraud Section, stated that CGMP False Claims Act cases could be "the next hot thing." Brenda Sandburg, Are GMP Violations the "Next Hot Thing" in False Claims Act Litigation?, PINK SHEET, April 1, 2013, at 38.

# Chapter VI
# The Lawyer's Role in CGMP Counseling

FDA's standards for CGMP compliance require highly technical, science-based design, analysis and control throughout the product life cycle. Thus, it would seem that CGMP compliance falls within the realm of scientists, not lawyers. The purpose of this chapter is to explain the role of lawyers in CGMP counseling, and how the advice that we provide is distinct from recommendations by outside regulatory consultants.

In rendering advice on issues related to CGMP compliance, a lawyer must have a thorough knowledge of the FDCA, applicable FDA regulations and guidance documents. Moreover, lawyers providing advice on CGMP compliance must have experience with FDA inspections, and understand how to respond appropriately to FDA advisory and enforcement actions.[1]

Prior to an inspection, the lawyer can provide valuable advice regarding inspection management techniques. Ensuring that clients understand the importance of proper inspection management is critical. For example, clients should understand the scope of FDA's inspection authority and the legal significance of an inspection refusal. Clients should also understand the proper procedure for FDA's entry and notice of inspection, and be advised to contact outside counsel if those procedures are not followed.[2]

The lawyer should provide advice to the client regarding the need to be truthful in all cases. Personnel should not shade the truth, or deliberately omit facts in an attempt to mislead the investigator. It is also important for personnel to understand that speculation is not helpful, and is not necessarily truthful. Communications with FDA investigators should be based in facts, not on speculation or guessing. With respect to documents, the lawyer should advise clients not to alter or hide documents. The lawyer should also remind the client to provide precisely what the investigator has requested. The client is not well served by an attempt to be helpful by providing more information than FDA desires. For inspections conducted at facilities outside the United States, the lawyer should ensure that facility managers understand FDA's expectations, and how those expectations might differ from those of the local authorities.

While FDA's inspection authority under section 704 of the FDCA is quite extensive; on occasion, FDA investigators will request documents that fall outside the scope of FDA's inspection authority, in whole or in part. If there is uncertainty about whether to furnish a document, or if part of a document is subject to FDA's inspection authority and another part is not, the client should call outside counsel.

---

[1] Rule 1.1 of the American Bar Association MODEL RULES OF PROFESSIONAL CONDUCT states that "A lawyer shall provide competent representation to a client. Competent representation requires the legal knowledge, skill, thoroughness and preparation reasonably necessary for the representation." MODEL RULES OF PROF 'L CONDUCT R. 1.1 (2007).

[2] For example, law enforcement personnel who are conducting a criminal investigation will not provide a Form FDA-482, Notice of Inspection. It is important for outside counsel to be involved in order to ensure that the client manages these investigations appropriately.

Similarly, the CGMP lawyer can provide advice regarding photographs and affidavits. Note that this conversation should occur well before an inspection, so the client will have time to revise its inspection management procedure, but in the event that this has not occurred, the client should contact outside counsel.

Following an inspection, a lawyer with significant CGMP experience can provide much-needed assistance to a client responding to a Form FDA 483. Because the lawyer's stock-in-trade is effective communication and advocacy, the lawyer can help ensure that the response is well written, complete and supported by facts. As an objective third party, the lawyer is able to provide advice regarding the significance of specific observations and potential consequences if FDA's concerns are not appropriately addressed. On rare occasions, FDA investigators include observations regarding issues that are outside the scope of FDA's authority, e.g., issues that are subject to another agency's jurisdiction or are governed by state law. The CGMP lawyer is in the best position to assist in drafting the response to this type of observation. The CGMP lawyer can also assist in drafting responses to Warning Letters, preparation for Regulatory Meetings and responding to other types of actions related to CGMP compliance.

A lawyer who provides advice and counsel on CGMP matters must think strategically in order to best protect the clients' interests. This means that the lawyer must understand the regulatory environment and FDA's "current thinking" on a wide range of issues. It also means that the lawyer must know when to engage attorneys in other practice areas, such as products liability counsel or securities counsel. A lawyer who provides advice on CGMP matters also must consider how a client should communicate with customers and the public about manufacturing issues. These communications, if structured appropriately, assure the public and FDA that the firm understands FDA's concerns and is working diligently to address those concerns. By contrast, a poorly worded communication can exacerbate tensions with the agency and possibly serve as the basis for third-party litigation.[3]

Lawyers who practice in this area should also know when to retain technical consultants on behalf of a client. Technical expertise in the form of highly qualified and trained independent CGMP experts can play a significant role in addressing concerns raised by FDA during an inspection. Through audits and remediation activities, technical experts can help avoid an FDA enforcement action or substantially curtail the length of any voluntary or court-ordered suspension of operations.[4] The selection of these independent experts is vitally important. Lawyers who have demonstrated skill and expertise in this area often have a better understanding of

---

[3] Rule 2.1 of the Model Rules states that "[i]n representing a client, a lawyer shall exercise independent professional judgment and render candid advice. In rendering advice, a lawyer may refer not only to law but to other considerations such as moral, economic, social and political factors, that may be relevant to the client's situation. MODEL RULES OF PROF 'L CONDUCT R. 2.1 (2007).

[4] As discussed previously, investigations conducted by counsel that would not be conducted in the ordinary course of business for the purpose of seeking legal advice is privileged. Moreover, although an attorney may be hired to represent a corporate client, the attorney must conduct its investigation through individual employees at the company. These conversations can be privileged attorney-client communications, but the attorney must be mindful of the scope of the privilege and how it can be destroyed. To determine whether a communication with an employee in the course of an attorney's investigation is privileged, the court considers four factors: 1) whether the communication was provided by corporate employees to counsel acting as counsel; 2) whether the purpose of the communication was to provide legal advice; 3) whether the non-attorney was aware that it was communicating to an attorney for the purpose of securing legal advice; and 4) whether the communications were confidential. *See* Upjohn Co. v. United States, 449 U.S. 383, 394 (1981).

It is critical that the lawyer provide the employee being interviewed with a warning, commonly referred to as the Upjohn warning, which informs the employee that 1) the lawyer represents the company, not any individual; 2) the interview is being conducted to assist in providing legal representation to the company; 3) the employee is not allowed to discuss the interview with anyone, including family and co-workers, absent a waiver of privilege; 4) only the company can waive the privilege; and 5) the interview will be disclosed to the company. The Upjohn warning is important because it notifies the employee that corporate counsel is not representing the employee in his or her individual capacity and clarifies the attorney's duty of loyalty to the company.

the reputations and qualifications of these experts and can play an important role in identifying, vetting, and selection of outside CGMP experts.

Outside counsel also typically plays the lead role in negotiating the terms of a consent decree of injunction focused on manufacturing and quality practices. Among the most important provisions that are often the subject of substantial negotiation with FDA are those that define the scope of the independent expert audit required by the consent decree. Because the person most familiar with the terms of the consent decree is the outside counsel who negotiated those terms, it is essential that the outside counsel play a significant role in reviewing and negotiating the terms of the agreement between the outside independent expert and the manufacturer to ensure that there is complete alignment with the scope of the audits required under the decree. Occasionally, there are disputes between the independent expert and the client about the scope of the audit. Outside counsel is usually best positioned to arbitrate those disagreements in a manner that is in the best interests of the client.

Outside counsel plays a key role in keeping the client apprised of new developments related to CGMPs, assisting in identification of risks, and development of risk mitigation strategies to help clients navigate CGMP compliance.

www.ingramcontent.com/pod-product-compliance
Lightning Source LLC
LaVergne TN
LVHW082355200926
848481LV00012B/3571

# PART I

# Conceptualising unconscionability